北方主要作物碳氮关系
对气候变化的响应与适应

宗毓铮　著

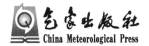

图书在版编目（CIP）数据

北方主要作物碳氮关系对气候变化的响应与适应 / 宗毓铮著. -- 北京：气象出版社，2021.1
ISBN 978-7-5029-7583-8

Ⅰ．①北… Ⅱ．①宗… Ⅲ．①气候变化－影响－作物－研究－中国 Ⅳ．①S5

中国版本图书馆CIP数据核字(2021)第211071号

北方主要作物碳氮关系对气候变化的响应与适应
Beifang Zhuyao Zuowu Tandan Guanxi Dui Qihou Bianhua De Xiangying Yu Shiying

出版发行：气象出版社
地　　址：北京市海淀区中关村南大街46号　　邮政编码：100081
电　　话：010-68407112（总编室）　010-68408042（发行部）
网　　址：http://www.qxcbs.com　　E-mail：qxcbs@cma.gov.cn
责任编辑：王元庆　　　　　　　　　　　终　　审：吴晓鹏
责任校对：张硕杰　　　　　　　　　　　责任技编：赵相宁
封面设计：艺点设计
印　　刷：北京中石油彩色印刷有限责任公司

开　　本：889 mm×1194 mm　1/32　　印　　张：6.125
字　　数：171
版　　次：2021年1月第1版　　　　　　印　　次：2021年1月第1次印刷
定　　价：48.00元

本书如存在文字不清、漏印以及缺页、倒页、脱页等，请与本社发行部联系调换。

前　言

近几十年来,全球大气 CO_2 浓度持续升高已成为不争的事实,预计 21 世纪末将达到 $421\sim936\ \mu mol/mol$。大气 CO_2 浓度增加使我国气温上升幅度大于全球平均水平。气候变暖引起的蒸发加大,风速、大风日数增加,土壤表层水分散失亏损致使农田干旱加剧,原本生产水平低下、产量不稳的农业生产面临更大威胁。在一些地区,水分、养分亏缺和大气 CO_2 浓度升高的交互作用严重影响着农田生态系统的生产力和生产效率,给粮食安全带来极大挑战,因而成为现代农业研究热点之一。

作物的碳氮代谢及其分配相互联系、不可分割,其生物过程及外界环境调节共同决定着作物生产力和营养水平。本书试图从细胞、组织、器官和个体等水平,就主要气候变化因子调节下(CO_2 浓度、水分与氮素供应等)作物的碳氮关系进行论述。深入了解这一过程,有助于系统理解气候变化下区域作物生产力变化的原因与机制,为未来气候变化下实现粮食安全与人类健康提供科学依据。

本书是作者及作者所在山西农业大学有机旱作与栽培生理创新团队气候变化对作物影响及农业适应课题组近年来在我国北方唯一的 FACE 平台(北京昌平)、位于黄土高原东部的 OTC 平台(山西太谷),以及人工环境控制室等开展的相关研究总结而成。本书共分 4 部分,分别对我国北方主要作物小麦、玉米、大豆,以及谷子、糜子等杂粮作物的光合作用、产量形成、营养元素利用过程对未来大气 CO_2 浓度升高环境的响应特征进行了阐述,并分析了目前研究所存在的问题及未来展望。

本书成果的研究和出版得到了国家重点研发计划项目子课题"气候变化背景下麦田不同减排措施的减排机理研究"

(2017YFD0300202-5)、国家自然基金面上项目"大气 CO_2 浓度升高对大豆脂肪代谢的影响及其调控机制"(31871517)、"高 CO_2 浓度下受旱玉米碳氮分配调节器官生长的生理机制"(31370425)、教育部高等学校博士学科点专项科研基金"二氧化碳和水分交互作用下玉米碳氮分配调节器官生长的机制研究"(20130204110024)、国家自然基金青年项目:"大气 CO_2 浓度升高对大豆抗旱性的影响机制研究"(31601212)、山西省应用基础研究计划项目"高浓度 CO_2 与水分胁迫交互作用下小麦成苗的碳氮运转机制"(201601D021124)的资助。多年的研究中得到了中国农业科学院、中国科学院水利部水土保持研究所、山西农业大学等单位的支持。在编著过程中得到了中国科学院水利部水土保持研究所上官周平研究员,中国农业科学院环境与可持续发展研究所郭李萍研究员、韩雪副研究员,澳大利亚墨尔本大学 Shu kee Lam 博士,以及山西农业大学高志强教授、郝兴宇教授、李萍教授、孙敏教授、杜天庆教授、杨珍平教授、张东升副教授、王卫锋副教授等老师的支持与帮助。杨净、龚泽华、牛冰洁等博士研究生,常翠翠、刘昭霖、任宏芳、王璐、刘亚静、张小琴、张媛铃、冯雅楠、李阿立、游文萍、张涵青、李买金、张鑫琪、董李冰、侯永强、张鼎谞、孟敏、工晨光、贺薇、乔红刚、勾俊英、王娜、郭晓磊、梁煜、李炳言、贾斯淳、刘紫娟、袁蕊、张仟雨、胡晓雪等硕士研究生先后参与了本课题组相关研究,在此表示感谢!

  作物在未来气候变化下的响应与适应是一个复杂的过程,本书仅是基于目前的研究,对大气 CO_2 浓度升高下北方主要作物的生理生态响应特征进行了初步探讨。由于水平有限,书中疏漏与错误之处在所难免,恳请读者批评指正。

宗毓铮
2020 年 6 月于山西农业大学

目 录

前言

第一部分 大气 CO_2 浓度升高对作物生长影响的
研究进展 ·················· 1

参考文献 ·················· 24

第二部分 大气 CO_2 浓度升高对主要 C_3 豆科
作物的影响 ·················· 41

第 1 章 大气 CO_2 浓度升高对大豆元素吸收
利用的影响 ·················· 42

第 2 章 大气 CO_2 浓度升高与干旱对大豆光合
特性与产量的影响 ·················· 51

第 3 章 大气 CO_2 浓度升高对绿豆元素吸收
利用的影响 ·················· 61

第 4 章 大气 CO_2 浓度升高对绿豆光合特性
与产量的影响 ·················· 72

参考文献 ·················· 79

第三部分 大气 CO_2 浓度升高对北方主要 C_3 禾本科
作物的影响 ·················· 87

第 5 章 大气 CO_2 浓度升高与不同供氮水平对
小麦光合特征和氮分配的影响 ·················· 88

第 6 章　大气 CO_2 浓度升高与干旱交互对小麦
　　　　光合特征和产量的影响 ·················· 99
参考文献 ································ 106

**第四部分　大气 CO_2 浓度升高对北方主要 C_4 作物
　　　　　的影响** ·························· 112

第 7 章　大气 CO_2 浓度升高对氮胁迫下受旱
　　　　玉米光合特征的影响 ·················· 113
第 8 章　大气 CO_2 浓度升高与氮胁迫对玉米幼苗干旱—
　　　　复水过程中光合特性的影响 ············ 125
第 9 章　大气 CO_2 浓度升高对糜子光合特性
　　　　与产量的影响 ························ 140
第 10 章　大气 CO_2 浓度升高对谷子光合特性
　　　　　与产量的影响 ······················ 150
第 11 章　大气 CO_2 浓度升高下青狗尾草与谷子
　　　　　光合特性比较研究 ·················· 168
参考文献 ································ 177

第一部分　大气 CO_2 浓度升高对作物生长影响的研究进展

自工业革命以来，人类大量燃烧化石燃料，向大气中排放了过量 CO_2 等温室气体，使全球大气 CO_2 浓度日益增高。根据政府间气候变化专门委员会(IPCC)报告(IPCC 2013)，全球大气 CO_2 浓度持续升高已成为不争的事实，预计 21 世纪末将达到 $421\sim936\ \mu mol/mol$ (Stocker, 2013; Feldman et al. 2015)。大气 CO_2 浓度升高会引发气温升高、干旱加剧、生物多样性锐减、传染疾病增多等严重后果，阻碍了生态系统的可持续发展，成为当前全球多个学科共同面临的重大课题(Reich et al., 2006)。稳定大气 CO_2 浓度的唯一途径是减少碳源并增加碳汇。减少碳源，是指在自然环境和人类活动中减少温室气体排放，但是会在短期内减缓经济增长。增加碳汇，是指通过增加植被覆盖，使植物将大气中的 CO_2 同化为有机物并释放氧气。因此，植物生长对大气 CO_2 浓度升高的响应备受研究者关注。

植物个体生长受基因控制，并随生存环境改变而调节各种代谢过程以完成生活史。近年来，大气 CO_2 浓度上升引起植物主要器官及不同发育阶段碳氮化合物分配和形态结构发生不同程度的变化(Westoby et al., 2002)。而器官中化学成分和化学计量比的变化影响其功能、生长和周转速率，并对植物生长做出反馈，这种反馈又进一步对植物的生活史对策、群落结构和进化策略，以及元素循环和物种共存产生重要影响(Niklas et al., 2002; Jackson et al., 1997; Urabe et al., 2002; Hessen et al., 2004)。另外，由大气 CO_2 浓度升高引起的降水格局改变(Houghton, 2001)，对部分地区水资源利用造成严重考验，使水分作为一种限制性资源影响着植物碳氮获取与分配。在我国,干旱与半干旱地区约占国土面积的52.8%,其降水量少而变率大,潜在蒸发量远大于降水量(山仑等,2002)。在旱地农业生态系统中,可持续发展的主要限制因子是降水少和由于土壤肥力低下引起的低水分利用率(山仑等,2002;高亚军等,2003)。

陆地生态系统中氮素和碳素的循环特性是衡量其稳定性的重要因素之一,它们在叶片、个体和生态系统水平上都紧密耦合(任书杰等,2006)。陆地生态系统中的氮主要通过影响有机碳的分解、植物的光合作用以及同化产物在植物器官中的分配等作用于碳循环过程(于贵瑞等,2011)。同时,碳循环过程为生物固氮、氮矿化、氮硝化和反硝化等提供了能量来源(Wright et al.,2004)。目前,关于气候变化下植物碳氮素交互作用的研究大部分集中于叶片和生态系统水平,关于个体尺度碳氮分配规律与形态构成的关系方面,还缺乏较为明确的认识。然而,个体尺度的生长调节和复杂的碳氮流交互作用直接或间接地影响着植物个体生长、种群生活史对策、群落结构和进化策略以及碳素和氮素的生物地球化学循环(Hessen et al.,2004),这是综合认识叶片尺度的光合调节和大尺度的生态系统适应关键的一步。所以,植株器官生长异质性、源库器官碳氮化合物分配、周转、再利用与形态构成的关系是阐明植物生长调节响应高浓度大气CO_2的重要切入点。这方面的工作也是目前植物形态学、植物生理生态学乃至气候学研究的重点领域。展开该方面的研究工作,对于深入认识植物化学计量调节和形态建设在高大气CO_2浓度下的响应机制、明确陆地植物群落对未来大气CO_2浓度升高的适应机理,以及探求提高植物水分利用效率的生理调控途径具有重要意义。

0.1 CO_2浓度升高对植物生产力的影响

0.1.1 对叶片光合作用的影响

由大气CO_2浓度升高引起植物叶片的光合适应目前已被研究者广泛关注(Gutierrez et al.,2009;Martinez-Carrasco et al.,2005;

Sarker et al. 2011)。在光反应过程中,高浓度 CO_2 可以通过增加单位叶面积叶绿素含量而提高叶片对光能的吸收能力与原初光能转化效率(张其德等,1997;林世青等,1996),同时使两个光系统之间激发能的分配格局发生改变(张其德 et al.,1997;林世青 et al.,1996),进而引发碳反应过程的不同适应。但也有报道与此结论相反,高浓度 CO_2 减少了单位叶面积叶绿素含量(卢从明等,1996),并未提高光能吸收能力。在碳反应过程中,不同光合途径植物对高浓度 CO_2 适应的敏感度存在差异。C_3 植物通常可以通过减小气孔开度、提高水分利用效率等途径在短期内维持光合速率,长期适应后下调光合性能(Long et al.,2004;Ainsworth et al.,2007;Albert et al.,2011)。例如,水稻幼苗光合速率在高浓度 CO_2 下处理 1 天显著增强,处理 7 天之后则显著降低,且 Rubisco 的表达受到抑制(唐如航等,1998),被称为光合适应或驯化现象。C_4 植物在高 CO_2 浓度下敏感性较弱(Wand et al.,1999;Sage et al.,2003;Ghannoum et al.,2011;Poorter,1993)。这主要是因为 C_4 植物叶片内存在叶肉细胞和维管束鞘细胞两种光合细胞,导致 C_4 植物拥有较高的 CO_2 浓缩机制(Ghannoum et al.,2011)。CO_2 最初先在叶肉细胞中被 PEP 羧化酶固定,随后扩散进入维管束鞘细胞,为 Rubisco 提供羧化底物(Ghannoum et al.,2011),使 C_4 植物的光合速率接近饱和(Ghannoum et al.,2011;2000)。因此,理论上空气中 CO_2 浓度升高对 C_4 植物的光合能力提高甚微。此外,光合作用对高浓度 CO_2 的适应与氮素水平有密切联系。研究普遍认为,高浓度 CO_2 使植株单位叶面积氮素含量减少,进而引起 C_3 植物的光合能力下调(Stitt,1991;Zhou et al.,2009)。在 C_4 植物中,由于叶片可以利用少量的 Rubisco 达到较高的光合效率,所以叶片氮含量的减少对其光合能力的影响不大(Furbank et al.,1996;Ghannoum et al.,2000)。

0.1.2 对植株生长的影响

近 40 多年来,研究人员通过室内控制实验与自由式 CO_2 气肥的小区实验(Free-air CO_2 enrichment,FACE)为研究高浓度 CO_2 对植株生长的影响累积了丰富资料(表0-1)。在瑞士连续10年的黑麦草试验表明,在氮肥充足的土壤中,CO_2浓度升高使黑麦草的产量显著升高,在氮肥匮乏的土壤中,CO_2浓度升高对其产量无显著影响(Schneider et al.,2004);然而,对加利福尼亚一年生草地的研究表明,增施氮肥不会显著提高其在CO_2浓度升高环境下的生物量(Dukes et al.,2005);对美国中北部的多年生牧草研究表明,在CO_2浓度升高的前3年,增施氮肥对牧草地上部生物量积累影响不显著,而从第4年开始,增施氮肥可显著增加牧草地上部生物量的积累(Reich et al.,2006a)。在我国北京、江苏等地也分别通过FACE试验进行了植物生长对高浓度CO_2的适应性研究,例如,在中国农业科学院运行的麦—豆轮作FACE实验表明,CO_2浓度升高主要通过降低不孕小穗数,提高穗粒数以提高冬小麦产量,氮肥供应不足会降低产量的提高幅度(韩雪等,2012);中国科学院南京土壤研究所对水稻/小麦生态系统连续10年的观测结果表明,短期内CO_2浓度升高可以促进土壤中阳离子的释放,并提高作物产量;但长期适应后,土壤酸性增强,对生态系统生产力的持续提高产生负面影响(Cheng L et al.,2010)。

表0-1 高浓度CO_2对植株生物量积累影响的长期试验

地点	年份	实验对象	CO_2浓度(ppm*)	交互处理	文献来源
北美洲					
美国(亚利桑那州马里科帕)	1989—1997	棉花、小麦、高粱	+200	供水条件、供氮水平	Ainsworth et al.,2005
美国(明尼苏达州)	1998—2006	多年生牧草	560	供氮水平	Reich et al.,2001;Reich et al.,2006a
美国(伊利诺伊州香槟市)	2001—2017	大豆、玉米、木薯、烟草	550、585、600	品种、供氮水平、增温	Bishop et al.,2015;Ursula et al.,2015;Leakey et al.,2006;Ruiz-Vera et al.,2020;Sanz-Sáez et al.,2017

* ppm为浓度单位,1ppm表示溶质质量为总质量的百万分之一。

第一部分 大气 CO_2 浓度升高对作物生长影响的研究进展

续表

地点	年份	实验对象	CO_2浓度 ppm	交互处理	文献来源
南美洲					
巴西 (Jaguariúna)	2011—2015	咖啡	+200	品种	Ghini et al., 2015
欧洲					
意大利 (Rapolano)	1995—1999	土豆、葡萄	+200、550		Miglietta et al., 1998; Magliulo et al., 2003; Bindi et al., 2001
德国 (Lindau)	1993—2004	黑麦草	600	供氮水平	Schneider et al., 2004
德国 (Braunschweig)	2000—2015	大麦、甜菜、冬小麦、玉米	550、600	供水条件、供氮水平	Weigel et al., 2012; Manderscheid et al., 2018; Dier et al., 2018
德国 (斯图加特)	2004—2008	小麦	+150		Högy et al., 2009 Högy et al., 2010 Högy et al., 2013
亚洲					
日本 (Shizukuishi)	1998—2008	水稻	+200	供氮水平、品种	Shimono et al., 2009; Hasegawa et al., 2013
日本 (Tsukubamirai)	2010—2014	水稻	+200	品种、增温、供氮水平	Sakai et al., 2019 Hasegawa et al., 2019
中国 (江苏无锡)	2001—2003	水稻	+200	供氮水平	Yang et al., 2006
中国 (江苏扬州)	2004—2008	水稻、小麦	+200	供氮水平	Yang et al., 2006; Zhu et al., 2009
中国 (江苏常熟)	2013—2014	水稻、小麦	500	增温	Cai et al., 2016

续表

地点	年份	实验对象	CO$_2$浓度 ppm	交互处理	文献来源
中国（北京）	2007—2020	小麦、大豆、绿豆	550	品种、供氮水平	Gao et al.,2015；Hao et al.,2016
印度（新德里）	2006—2012	绿豆、马铃薯、花生、水稻、小麦、芥菜、鹰嘴豆	550		Singh et al.,2013
大洋洲					
澳大利亚（维多利亚州霍舌姆）	2007—2016	小麦、小扁豆、紫花豌豆、蚕豆	550	品种、供水条件、热浪	Fitzgerald et al.,2016；Macabuhay et al.,2018；Bourgault et al.,2016；Parvin et al.,2019；Bourgault et al.,2018
澳大利亚（Walpeup,VA）	2008—2009	小麦	550		Fitzgerald et al.,2016

植株生物量的积累与新器官生长密不可分，新器官的生长依赖于源器官光合产物的供给与库器官对同化物的利用（Wardlaw,1969）。一方面，CO$_2$浓度升高导致叶片气孔开度减小，蒸腾作用下降，水分利用率提高，光合产物增加（Martin et al.,2007；Casson et al.,2008），但不同光合途径植物的干物质增加幅度存在差异，C$_3$植物的生物量将平均提高41%，C$_4$植物平均提高22%，CAM植物平均提高15%（Poorter 1993）；另一方面，长期定位试验证明，高CO$_2$浓度下C$_3$植物的库器官活力下调（Reich et al.,2006；Ainsworth et al.,2010），源叶糖分含量升高，抑制光合相关酶类的基因表达，导致源器官光合活性下降，库活力降低（Leakey et al.,2009），产量下调（Drake et al.,1997；Long et al.,2006）。所以，在未来大气CO$_2$浓度升高条件下，加强碳素利用，光合产物的运输是提高光合作用与产量的重要因素。

0.2 植物体内碳氮素积累与分配机制

0.2.1 植物对碳氮素获取与利用的一般模式

碳氮营养是植物生长所必需的大量元素。一般来讲,植物生长的碳氮营养供应可分为体外和体内两个来源,体外氮素供应有 4 种来源:土壤有机质(或肥料)的矿化、微生物对空气中氮气的固定、根际共同体所转化的有机氮或是氮沉降(上官周平等,2004);体外碳素供应主要通过叶片对空气中 CO_2 的光合同化作用,植株体内的碳氮源均来自通过生理过程储存在体内的物质再代谢及循环(Millard et al.,2010)。

植物体中碳素和氮素的储存通常存在 3 种形式:当供应超出需求时,植物为持续生长和繁殖所进行的积累与为生长或防御而形成的储备,例如:坚果形成时的籽粒储备,以及为代谢周转而进行的再循环,例如叶片衰老时的蛋白转化(Chapin Ⅲ et al.,1990)。碳氮素在体内的循环过程如图 0-1 所示,虚线箭头表示体外与体内碳氮库间的资源转化途径:植物体吸收的碳氮同化物用于生长、繁殖及其他代谢;吸收后的物质在储存库中积累或用于的再运转;为生长、防御、竞争而形成的储备;体内碳氮素通过叶片和根系衰老脱落而损失。实线箭头表示体内碳氮循环途径:储存库中积累的同化物由于季节性需求而再转运,以满足其余植物组织的生长代谢需求,当叶片衰老或优先脱落时,组织中的营养物质被运出,通过再循环而被重新储存(Millard et al.,2010)。如果植物体中积累的碳氮化合物不可移动也不可被再利用,就认为被植物固定。

0.2.2 碳积累与分配

碳分配是指植被初级生产力(gross primary production,GPP)以碳的形式向根、茎和叶等器官运输的过程,以用于呼吸和初级、次级产物合成。在秒或小时尺度上,光合作用生成的碳水化合物在植

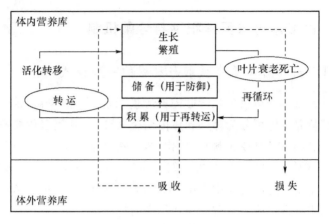

图 0-1 植物体对碳氮的获取及体内循环
(Millard et al.,2010)

物体内的分配(carbohydrate allocation)包括蔗糖、淀粉和非结构性碳水化合物,这种分配依赖于当前碳同化速率;在月或年尺度上,各器官光合产物的积累包括为生长或防御而储备的非结构性碳水化合物(Tilman,1989;Dewar,1993)。非结构性碳水化合物的积累遍布植物整体,包括叶片、茎秆和根系(Hoch et al.,2003;Würth et al.,2005;Palacio et al.,2008;Spann et al.,2008),其积累量与季节变化和植物生活型有关(Sauter et al.,1994;Hoch et al.,2003;Spann et al.,2008)。一些研究表明,植株可通过多种方式避免在快速生长和遭受侵扰时碳储存利用枯竭(Palacio et al.,2008;Hoch et al.,2003)。

光合作用决定了叶片中可利用碳素的总量,而可以用来分配的碳量依赖于体内代谢能力。光合产物在体内的分配受多种因素控制,包括光合速率、呼吸速率、器官活力与同化物传输储存策略,而环境因子与器官源库功能的变化进一步增加了同化物的分配过程的复杂性(Taiz et al.,2006;平晓燕等,2010)。目前,关于同化物分配机制的假说主要有 3 种,即功能平衡假说、源汇关系假说与相关生长关系假说。3 种假说自身均具有优点与缺点,各学派之间仍存

第一部分 大气 CO_2 浓度升高对作物生长影响的研究进展

在一定分歧。

功能平衡假说(functional equilibrium hypothesis)认为,植物通过冠层与根系之间相互制约与促进决定个体生长和同化物的运输方向,进而保证资源最优化利用(平晓燕等,2010)。当冠层光合能力强时,向根部供给的同化物可以保证根系生理活动正常进行,则促进根系生长,反之则抑制;当根部活力较高时,向冠层提供的水分与营养物质可以满足冠层生理活动与营养物质储存的需求,则促进冠层生长,反之则抑制。该假说可以概述植株营养生长阶段的同化物分配特征,已有不少生态学家利用其解释大气 CO_2 浓度升高对草地与森林生态系统碳分配的影响,大部分研究表明 CO_2 浓度升高会增加同化物向地下部的分配比例(Cotrufo et al.,1997;Ge et al.,2012;Iversen et al.,2008)。但是,功能平衡假说对植物生殖生长阶段分配规律缺乏进一步认识。

相关生长关系假说(allometric relationship)用相关方程表示了植株个体发育与生长速率之间的关系,同时反映了器官组成结构与生理过程对个体大小的影响(平晓燕等,2010)。例如,树木的胸径与树高间的相关生长关系(樊艳文等,2011);小麦穗分化进程、叶面积指数、源库指标等与生长的相关关系模拟(薛香等,2006;朱艳等,2004),这些模型可以通过改变相关方程中的系数与指数来区分同化物分配的转变原因是外部环境因素还是内部生长发育(Cheng et al.,2007;McCarthy et al.,2007)。运用该假说对植株个体生长趋势的模拟还有助于反应群落甚至更大尺度的生态学特征(Wenxuan et al.,2003),但不能在机理上揭示同化物在不同器官之间的分配规律。

源汇关系假说(source-sink relationship hypothesis)强调源器官与库器官在个体生长中的作用,认为个体生长是由源器官(功能叶片)与库器官(新生叶、茎、果实及根)之间动态调节的整体。源器官通过韧皮部输送同化物形成"流",库器官之间竞争得到同化物(Taiz et al.,2006)。同化物的分配就是由源、库及流三者之间相互作用达到动态平衡。该假说符合植株解剖和发育模式(Taiz et al.,

2006),能较好地解释个体发育过程各个器官功能转变对同化物分配的影响。例如,在小麦幼苗期,成熟叶片是源器官,幼叶是库器官,随着植株生长发育,幼叶逐渐成熟,其自身结构与功能发生变化,导致同化物运输方向从输入变为输出,从而具备了源器官的功能,而旧的成熟叶则老化变黄。源汇关系假说是一个半机理假说,已有不少研究利用源汇关系假说研究体内同化物分配策略对环境变化的响应。Cheng 等(2010)利用稳定同位素示踪的方法研究夜间高温条件下水稻体内同化物源库转运特征,表明库器官籽粒中同化物分配减少是水稻产量降低的主要原因(Cheng et al.,2010)。Ruehr 等(2009)利用同样的方法证明干旱减缓同化物在源叶中的输出速度与根系中的输入速度,进而导致土壤物质循环发生改变。

0.2.3 氮积累与分配

无机氮被吸收后在植株体内还原、运输、合成、转化及再运转至各个部位,与蛋白质代谢共同构成生命活动的基本过程(Taiz et al.,2006)。在植物体内,氮素以多种营养蛋白的形式储存,例如代谢活性蛋白 Rubisco(Millard et al.,2007)。其中,直接用于生长和代谢的氮素在植株地上部的分配随生长中心和光在冠层中分布而转移。在生长初期和中期,植物生长速度较快,冠层顶部的高光强要合成较多的光合产物,需要大量光合蛋白供应,对叶片中氮素运输形成了一种"拉动"作用,促使下层叶片中的氮素上移至上层叶片(Anten et al.,1995),用于生长或繁殖(王忠强等,2007)。生长末期叶片开始衰老,蛋白质逐步分解或转移向其他植物组织以备再用(Silla et al.,2003),叶片光合能力下降,对氮素的需求减少,因而氮储存库转移向根(范志强等,2004),只有 10%～20%的氮存留在老叶中,最后以凋落物的形式将这部分氮归还到土壤中。氮素在冠层的分布还与冠层结构特征、叶片生理年龄和氮素供应量等有关。随着叶片生理年龄的衰老,叶片氮素垂直分布梯度变小(Gastal et al., 2002)。氮素供应充足时,叶片氮素分布梯度不会因叶片衰老而发生改变(Hikosaka et al.,1994)。也有研究认为,氮素供应量对冠层顶部的氮素含量影响不大(Shiraiwa et al.,1993;Dreccer et al.,2000),或使其分布更加均匀(Milroy et al.,2001),但能较大地改变叶面积指数与单数之间的回归函数关系(Gastal et al.,2002)。

与非结构性碳水化合物库相反的是,在月或年尺度上,可以被再运转的氮素储存不会遍布于植物整体。不同生活型植物氮素储存于不同的器官中,落叶树种大多将氮储存于根和茎的木质部和韧皮部(Millard et al.,1989),当植物体氮素亏缺时,所有的氮储存库都可能被调动甚至耗尽,如衰老叶片中 Rubisco 的选择性损失(Millard et al.,1989);常绿针叶树种的氮素主要储存于新生针叶的光合蛋白(Nambiar et al.,1987;Millard et al.,1992;2001)。研究指出,针叶中有两个可被独立控制的 Rubisco 库,其储存量在秋季远超过光合利用需求(Millard et al.,2007),以满足春季雨水冲刷后所需的氮素调运(Manter et al.,2005)。

植株体对碳和氮的再利用以及植株的生长速度主要由氮储存情况决定,而不是碳水化合物(Cheng et al.,2002)。氮的存储和代谢能力与生长季节紧密相连,不受当前外部供应量影响。植株在生长和防御中所需氮素通过对氮库中氮素的再代利用和外界氮素的直接吸收得到满足。两者的平衡决定了植物氮素的损失与利用效率之间的比率(Millard et al.,2010),因此环境中的氮库利用也对氮平衡产生重大影响(表 0-2)。

表 0-2 植物碳氮储存与再利用的生理特征(Millard et al.,2010)

项目	氮素	碳素
储存	冠层衰老或休眠时仍可进行	由光合速率决定
	与物候相关,具有季节性	以碳固定的形式避免光合下调
	储存于特殊组织与器官	储存遍布整体植株
再利用	原有存储被耗尽	原有存储未被耗尽
	由源驱动	由库驱动
	不受当前氮素供应条件影响	受当前光合速率影响

0.2.4 碳氮的耦合

碳素与氮素的循环是陆地生态系统最重要的物质循环过程,伴随着能量传递与养分循环(于贵瑞等,2011)。从植物叶片尺度、个体尺度、生态系统尺度、区域尺度到全球尺度,碳循环与氮循环之间存在多个层面的耦合机制,这些耦合机制是了解陆地生态系统碳氮循环过程的理论基础,其生物过程的偶联作用与外界环境调节共同决定着植物生产力。

从生态系统与区域尺度来讲,陆地生态系统光合物质生产和碳积累过程与土壤氮供给能力的平衡关系是制约生态系统生产力的主要因素。光合作用是陆地生态系统碳固定的主要途径,氮素的吸收与转运为这一过程的进行提供物质基础(Millard et al.,2010),所以土壤氮素供给能力直接影响着生态系统碳固定能力与积累潜力。有研究表明,在高浓度 CO_2 环境中,一些植物碳同化能力增强,生物量积累加快,对氮素供给提出了更高需求,在氮素匮乏地区,氮素供给的有效性低是限制生态系统净初级生产力的主要环境因子(Reich et al.,2006b)。

从植物个体尺度来讲,碳氮资源分配与转运格局是一个平衡的动态变化过程。营养物质根冠周转速率是决定植株生长力的基本要素,并对生长过程产生直接影响。植物光合能力与光合器官中的氮含量相关性较高(Gastal et al.,2006),叶片中光合作用酶(特别是羧化酶)的数量与活性以及氮素在光合作用酶的分配比例均一定程度上调节光合能力(Wong,1979),而光合器官中的氮素又依赖于根系对氮的吸收与运输,根系中硝酸盐的还原与铵盐的积累需要利用光合作用提供的碳基础性物质作为能量和碳骨架(Millard et al.,2010;Gastal et al.,2002)。

关于个体碳—氮化学计量有一些经典研究,例如 Mooney 和 Gulmon 设计了一系列计量模型,研究环境因素变化与光合作用相关酶浓度的计量关系。研究表明,叶片氮输入提高了光合作用酶水平,在计算冠层光合固碳所获得的能量时必须减去叶片氮输入所消耗的能量(Mooney et al.,1979)。Field 研究了叶片氮输入与输出所消耗的能量关系,指出单个叶片氮输入所消耗的能量等于氮输出叶片降低光合能力所损失的能量,如蛋白降解、氨基酸运输和蛋白再合成造成的能量损失(Field et al.,1986)。植株通过碳氮循环调节不同代谢过程以适应生长需求,这些过程需要考虑碳氮吸收和同化、碳氮化合物的运输和分配、器官呼吸、新组织生长对碳氮化合物的利用、碳氮储存,以及根系分泌。已有较多研究表明,在高浓度 CO_2 环境中,植株生长速度加快,向地下部投入较多碳素,减少叶片单位生物量氮含量,改变体内原有的碳氮平衡(Dyckmans et al.,2005;Xu et al.,2007),新的碳氮分配格局与转运机制亟待研究。

0.3 高浓度 CO_2 改变植株体内碳氮分配

大气 CO_2 浓度的升高会引起植物细胞生物化学组成、体内碳氮周转和分配格局的改变。低化学计量过程的变化可以控制高生态学过程格局,所以这种变化直接或间接地影响着植物个体生长、种群生活史对策、群落结构和进化策略以及碳素和氮素的生物地球化学循环。

0.3.1 叶片尺度

营养物质的同化过程就是将稳定的低能量的无机化合物转变成高能量的有机化合物。有研究表明,一颗植物要用它 25%的能量来转化不足其干重 2%的氮(Taiz et al.,2006)。许多这种同化反应是在拥有强有力还原剂的叶绿体基质中进行的,如在光合作用电子传递中产生的 NADPH、硫氧还蛋白和铁氧还蛋白。营养物质同化和光合作用电子传递相偶联的过程被称为光合同化作用(Taiz et al.,2006)。虽然光合电子传递与卡尔文循环发生在同一部位,但是只有当光合电子传递过程产生的还原剂剂量超过卡尔文循环的需要量时(如强光低 CO_2 条件)光合作用才能进行(Robinson,1988)。

CO_2 浓度升高后,蔗糖在库器官中的浓度升高会促进植物生长基因表达(Farrar et al.,1996),但如果氮或其他环境因素受到限制,碳与氮的交互作用会引起光合速率降低,限制产量持续增长。原因可能有3种,首先,叶片的光合能力与Rubisco数量直接相关,高浓度CO_2下Rubisco加氧反应被抑制,使较低水平的Rubisco维持一定水平的光合速率,从而只有较少的氮分配到光合作用酶上,而较多的氮被分配到植物其他组织中而不是叶片(图0-2)(Tingey et al.,2003);其次,当植物对碳的净吸收量超出碳水化合物的运输和利用能力,蔗糖在叶中的积累抑制Rubisco表达(Sheen,1994),进而增加了蔗糖在叶肉细胞的循环(Ainsworth et al.,2005)。这两个理论被称为多源限制(multiple resource limitation)(Reich et al.,2006b)。第三,高浓度的CO_2抑制C_3植物叶片中和C_4植物维管束鞘细胞中硝酸盐的同化(Becker et al.,1993),而根系对氮的吸收能力也受根系中碳可溶性化合物浓度和韧皮部中氮可溶性化合物浓度的负反馈作用调节(Gastal et al.,2006)。所以,当高CO_2降低了土壤氮利用率或增加生物量中碳的含量,就会降低叶片氮含量,抑制植株体内Rubisco的数量和活性,进而增加碳水化合物的积累,降低光合能力(Ellsworth et al.,2004;Nowak et al.,2004)。

0.3.2 个体尺度

在个体尺度上,由CO_2浓度增加引起的氮限制作用主要表现在降低叶片氮含量,影响净光合速率;改变植物化学成分和化学计量比,对草食动物、疾病和真菌共生造成影响;改变生物化学循环;影响植物—土壤微生物系统的组成;增加氮固定或固定器数量(Reich et al.,2006b)。原因主要在于高浓度CO_2对同化物分配的影响受体内氮素和非结构性碳水化合物含量的共同控制(Poorter et al.,2000;Andrews et al.,2010),CO_2浓度升高使体内氮含量降低(Reich et al.,2006b;Zak et al.,2000),根系C/N(Xu et al.,2007;Johnson et al.,2001;PA,2001)和根分泌物增加,以便吸收更多的氮来保证养分供应。

在贫瘠地区,土壤中可有效利用的氮素含量低,增加光合产物在根系中的分配比例会加重土壤氮限制,进而限制CO_2气肥作用,

第一部分 大气 CO_2 浓度升高对作物生长影响的研究进展

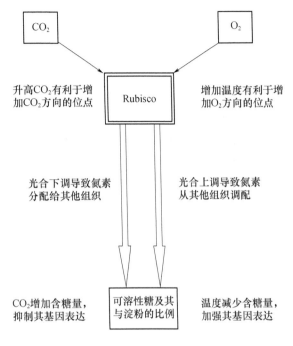

图 0-2 CO_2 浓度升高条件下 Rubisco 的调节框图
(许振柱等，2007)

这个过程被称作渐进式氮限制(progressive nitrogen limitation, PNL)(Luo et al. ,2004)。渐进式氮限制通过减少植物氮利用率,降低生物量中碳化合物的含量,限制植物生长,甚至在氮肥供应不受限制时也可以发生(Reich et al. ,2006b)。多源限制与渐进式氮限制共同作用,加重高 CO_2 对氮贫瘠地区氮肥有效利用率的限制;同时,CO_2 浓度升高会增加 C_3 草本植物源叶的淀粉含量,使叶片和根系之间产生膨压梯度,导致相对较多的光合产物向根系分配(Farrar et al. ,2006;Suter et al. ,2002),这种影响在养分匮乏下较为明显。但也有研究表明,CO_2 浓度升高对植物光合产物分配格局影响变化趋势不一,谷物根冠比保持平衡(Farrar et al. ,2006),柠条增加(Xu et al. ,2007),树木根冠比下降(Farrar et al. ,2006)或升高(Zak et al.,2000;Liu et al.,2002)结果不同,这与物种生理特性及植物生长条件的控制,如土壤水分、养分浓度、生长温度、生长季长短、CO_2处理时间长短等都有相关。在土壤氮供应充足条件下,CO_2浓度升高引起的根系C/N和根分泌物的增加可以加速土壤—植物营养循环,促进植物生长,长期胁迫不改变体内光合产物分配格局(Liu et al.,2002)。然而,高浓度CO_2可以影响土壤氮循环过程,调节植物氮利用效率。这个过程偶联的化学计量改变也可以削弱生态系统碳获取及储存的能力(Reich et al.,2006b)。

大气CO_2浓度增加改变植物化学成分和化学计量比,进而改变生物化学循环,影响植物—土壤微生物系统的组成,改变氮固定和固定器的数量(Reich et al.,2006b)。虽然目前许多关于个体生长的研究涉及器官水平,但是从植株整体发育的角度对复杂交错的碳氮化合物分配调节规律依然没有明确的解释。揭示这一复杂规律有助于理解体内交互复杂的物质流系统,预测气候变化对旱地植物生产力的影响。

0.4 大气 CO_2 浓度升高与其他环境因子的交互作用

0.4.1 氮素与大气 CO_2 浓度升高的交互作用

氮素供应不足可能会限制大气 CO_2 升高对 C_3 植物生长的促进作用(Sakai et al.,2019;Hasegawa et al.,2019)。在全球不同区域的 6 个 FACE 实验站的 25 项研究表明,在大气 CO_2 浓度升高条件下,低氮供应的 C_3 作物产量较正常供氮平均降低了 8%(Sakai et al.,2019;Hasegawa et al.,2019)。但也有研究表明,CO_2 与氮素交互下作物产量较对照没有变化(Weigel et al.,2012;Yang et al.,2006)。比如,位于德国的 CO_2 与施氮(常规与常规施氮量的 50%)交互作用的定位研究表明,在大麦、黑麦草、甜菜、冬小麦的轮作中,大麦、黑麦草、甜菜的产量均没有显著变化,仅冬小麦的产量呈降低趋势(Weigel et al.,2012)。在 CO_2 浓度与硝酸铵钙肥(40,180,320

kg N/ha)的交互试验中,最低供氮水平限制了 CO_2 浓度升高下小麦地上部生物量的积累,但是中氮与高氮下均表现为促进(Dier et al., 2018)。杂交稻的产量在低氮下也没有表现出明显降低,且大气 CO_2 浓度升高下吸氮量在与两种施氮水平下均提高10%,与产量的提升结果基本一致(Yang et al., 2006)。大气 CO_2 浓度升高条件下,植物对氮素限制的响应存在较大差异的主要原因有:土壤本底肥力、气候条件与作物管理措施不同等。所以,在全球不同区域开展多点试验才能深入了解不同气候与土壤环境下 CO_2 与氮素交互对作物生产力的影响。

前人研究表明,大气 CO_2 浓度升高下作物光合能力提高,体内碳水化合物含量增加,但根系对矿质元素的吸收量没有明显变化,进而改变了植物体内原有的化学计量平衡,降低了籽粒中矿质元素含量,即碳水化合物对矿质元素的稀释作用(Loladze,2002)。所以,近年来关于未来气候变化环境下人类对必须元素摄取不足的问题受到越来越多的研究人员关注(Loladze, 2014; Myers et al., 2014; Zhu et al., 2018)。不少研究认为,大气 CO_2 浓度升高下作物籽粒蛋白质与氮含量降低的主要原因包括:(1)稀释作用(Loladze 2002);(2)凋落物体内 C/N 高,分解速度下降(Uddling et al., 2018);(3)土壤微生物氮有效性的降低(Uddling et al., 2018);(4)地上部硝酸盐同化的限制(Uddling et al., 2018);(5)光合作用过程中 Rubisco 需求量的减少(Ainsworth et al., 2005; Wang et al., 2020; Smith et al., 2020);(6)器官分配格局发生改变;(7)蒸腾降低引起的木质部物质运输速度减缓(McGrath et al., 2013)。其中,稀释作物是籽粒蛋白质含量降低的主要原因(Taub et al., 2008; Taub et al., 2008)。

然而,豆科植物常常在大气 CO_2 浓度升高下表现出较好的适应性,主要原因可能是根系通过根瘤菌固氮过程以一定量的碳水化合物换取了更多的吸氮量(Ort et al., 2006)。FACE 条件下的大豆(Myers et al., 2014)、桃树(Myers et al., 2014)与小扁豆(Bourgault et al., 2018)籽粒蛋白质含量均未发生明显下降(Myers et al., 2014)。同时,丛枝菌根被认为在获取矿质营养(特别是磷)的过程中起到重要作用(Smith et al.,2012)。研究表明,水稻、桃树、大豆体内磷含量在大气 CO_2 升高下没有显著变化(Myers et al.,2014),而菌根生物量增加了约 7%(Dong et al.,2018)。因此,菌根的生物合成被认为是植物适应未来 CO_2 浓度下土壤环境的一个关键过程(Strullu-Derrien et al.,2018)。所以,缺乏丛枝菌根真菌结合能力的植物(如十字花科)在大气 CO_2 浓度升高环境中生长不占优势(Keymer et al.,2017)。目前,十字花科植物对大气 CO_2 浓度响应的研究较少。但在印度 FACE 条件下进行的研究表明,芥菜(*Brassica juncea*)产量在大气 CO_2 升高下增长了 20%,在德国的油菜产量增加了 18%(Högy et al.,2010;Singh et al.,2013)。这个增产幅度超过了关于小麦与玉米的报告,接近 C_3 作物的平均水平。

为证明 CO_2 升高环境下土壤中氮素有效性与硝酸盐同化限制对植株体内氮积累的抑制作用,在一项 FACE 研究中,前人设计了分别以铵态氮与硝态氮为主的两类氮源处理(Dier et al.,2018)。两种氮源条件下,地上部生产力没有显著差异,但是吸氮量在 CO_2 升高下差异显著,以硝酸盐为主要氮源的植株吸氮量在 CO_2 升高下显著高于以铵态氮为主要氮源的植株。这表明硝酸盐同化限制不是大气 CO_2 浓度升高下植物组织氮浓度降低的主要原因(Dier et al.,2018)。在水培控制试验中,当硝酸盐是主要氮源时,CO_2 升高限制了氮同化,说明硝酸盐可能会争夺还原剂。当以铵盐为主要氮源时,这种情况明显减少(Bloom et al.,2010)。

另外,蒸腾速率的降低被认为与大气 CO_2 升高下植株地上部元素含量降低有较大关系,但是在一些研究中这种关系较弱。在 SoyFACE 平台,大豆生长季蒸发散效率降低了 8.6%,但是体内 Fe 含量仅降低了 4.1%(Myers et al.,2014;Bernacchi et al.,2007)。在同一个 FACE 平台,玉米生长季蒸发散效率降低 9% 时,Zn 含量仅下降了 5.1%(Hussain et al.,2013;Myers et al.,2014)。增温可以提高蒸腾速率,在与 CO_2 升高的交互作用下,可以减轻 CO_2 升高下地上部元素含量降低幅度(Köhler et al.,2019)。然而,在位于德国的 FACE 研究中,蒸腾速率的降低并没有引起玉米与小麦体内锌、铁等多种矿质元素积累量的变化(Manderscheid et al.,2016; Manderscheid et al.,2014;Erbs et al.,2015)(Dier et al.,2020; Manderscheid et al.,2018)。甚至,在一项位于澳大利亚南部的 FACE研究中,小麦植株单位蒸腾量下,一部分元素在植株体内的积累量降低,但钙、镁与锰的积累量增加(Houshmandfar et al.,2018),表明可能存在补偿机制。目前,仍需要更多的研究证明 CO_2 浓度升高下地上部元素含量降低的机理。

0.4.2 干旱与大气 CO_2 浓度升高的交互作用

CO_2 浓度升高下,多数 C_3 作物具有两个基本适应特征:光合碳同化增强与气孔导度降低(Ainsworth et al.,2007;Long et al., 2004)。所以,在 CO_2 与干旱交互下,作物可能通过减少耗水,加快冠层生物量积累来增加产量(Bernacchi et al.,2015)。对位于干旱半干旱气候的美国亚利桑那州马里科帕的FACE试验平台多年研究结果的meta分析表明,棉花、小麦与高粱在大气 CO_2 升高下产量平均提高了28%(Ainsworth et al.,2005)。其中,棉花的增产幅度最大,在正常灌溉与轻度干旱条件下均达到了43%(Mauney et al., 1994),小麦产量在干旱条件下的提升幅度大于正常水分条件(Ainsworth et al.,2005),高粱仅在干旱条件下增产,在正常水分下增产不显著(Ottman et al.,2001)。但是,不同区域的FACE平台试验结果存在较大差异。在位于干旱与半干旱气候的澳大利亚Horsham与Walpeup的Grains FACE站进行了小麦与小扁豆研究,位于温带湿润大陆性气候的Champaign,Illinois进行了大豆研究。在这些FACE研究中,小麦、玉米与小扁豆的产量在湿润条件下的提高幅度均高于干旱条件(Fitzgerald et al.,2016;Gray et al., 2016;Houshmandfar et al.,2018;Parvin et al.,2018)。同时,在大气 CO_2 升高下早期生物量的快速积累与茎/根比的升高对后期生长具有一定副作用,尤其是在重度干旱条件下。对小麦的研究表明,干旱与 CO_2 交互条件下地上部生物量的增加导致有效分蘖减少、后期籽粒灌浆不充分(Houshmandfar et al.,2016)。在大气 CO_2 浓度升高下,干旱条件下的大豆与小扁豆生育前期冠层吸水量并没有受到气孔导度降低的影响(Gray et al.,2016;Parvin et al.,2018),但湿润条件下土壤含水量更高,而足够的土壤含水量是在大豆灌浆期植株维持继续固氮与产量增加(Parvin et al.,2018),以及籽粒蛋白质积累(Gray et al.,2013)的基础。

在干旱条件下,植物通过减少气孔导度(G_s)与冠层耗水来克服土壤含水量降低,导致大气 CO_2 浓度升高背景下干旱对作物的影响大于正常灌溉,使作物在干旱环境中维持较长的生长时间(Kimball 2016;Leakey et al.,2010;Leakey et al.,2009;Tausz-Posch et al.,2015)。另外,拟南芥与木豆的研究表明,在大气 CO_2 浓度升高环境中,二者具有较高的抗氧化能力,因而降低了光呼吸与氧化性损伤,提高了抗旱性(Zinta et al.,2014;Sreeharsha et al.,2019)。近年来,也有 FACE 平台的试验结果表明,C_3 作物对大气 CO_2 浓度升高的响应在湿润与中度干旱条件下没有明显差异。在美国亚利桑那州马里科帕的 FACE 研究表明,当蒸腾量减少到正常时的 50%~75% 时,部分 C_3 作物的产量可能会由于土壤含水量在雨季与旱季之间变化较大而降低。

多数研究表明,大气 CO_2 浓度升高可以减轻轻度与中度干旱胁迫,但不能减轻重度干旱胁迫(Gray et al.,2016;Sekhar et al.,2020)。比如枫香(Warren et al.,2011),大气 CO_2 浓度升高下,枫香的 G_s 与蒸腾速率(T_r)在中度干旱下下降幅度较大,水分吸收与运输能力降低,重度干旱下 G_s 进一步降低,抵消了 CO_2 的气肥作用,最大光合速率(A_{Sat})显著降低(Warren et al.,2011)。也有研究表明,对于部分树种,CO_2 浓度升高也不能减轻干旱胁迫,比如杨树与柳叶桉。因为杨树具有较大的叶面积、较高的气孔密度与低的木材密度,这些都使杨树对干旱胁迫较敏感(Bobich et al.,2010)。然而,柳叶桉在 CO_2 浓度升高下也没有明显的减轻干旱胁迫,因为较小的根系体积没有有效改变体内水分状况(Duursma et al.,2011)。所以,CO_2 气肥对作物生长的促进作用在干旱条件下被削弱了,特别是气孔关闭程度较高时(Warren et al.,2011)。

有研究认为,与 C_3 植物相比,C_4 植物叶片由于具有 CO_2 泵机制,在正常大气 CO_2 浓度下叶内 CO_2 传导与利用效率已经维持在较高水平,因而具有较高的光合能力(Ghannoum et al.,2000)。所以,在大气 CO_2 浓度升高条件下,C_4 植物可能缺乏促进 A_{Sat} 进一步提高的内在动力,产量提升幅度也较低(Leakey et al.,2006)。但是,也有研究认为,在干旱与 CO_2 交互条件下,C_4 植物光合能力的提升幅度接近于 C_3 植物(Meng et al.,2014)。

0.5 大气 CO_2 浓度升高对植物影响的研究方法

有关大气 CO_2 浓度升高对植物生理生态影响的研究,由于研究目的、试验水平、植物种类、生态系统类型等因素的差异,科学家可采用不同的研究方法进行材料培养。从1980年有关高浓度 CO_2 研究开始至今,主要的方法有控制环境试验(Controlled environment,CE)、开顶式同化箱(Open top chambers,OTC)和自由式 CO_2 气肥试验(Free air CO_2 enrichment,FACE)。全球生态学研究人员利用不同研究手段对高浓度 CO_2 下植株生长不同尺度的适应性研究积累了丰富资料。

0.5.1 控制环境试验

控制环境试验可以在温室、人工气候室或是生长箱中进行。这类试验主要通过在封闭的室内严格控制温度、湿度、光照和大气 CO_2 浓度等环境因素,以其中一种或几种因素作为变量达到试验需求。它的优点是可以人为组合多种环境因素,在一定的假设条件下进行重复试验,并对环境因子控制精度较高,被许多植物生理学家接受,如德国哥廷根大学 Dyckmans 等利用控制环境试验对山毛榉(*Fagus sylvatica* L.)碳氮分配模式在高浓度 CO_2 条件下的响应进行一系列研究(Dyckmans et al.,2005,2001;Dyckmans et al.,2000)。在国内,Zhou 与 Shangguan 利用同样方法研究 CO_2 浓度升高对红豆草(*Onobrychis viciaefolia* Scop.)幼苗氮素利用的影响,证明高浓度 CO_2 降低了单位生物量氮含量(Zhou et al.,2009)。但是,控制环境试验不能较系统地模拟外界环境变化,如风速、光照和湿度等的日变化和季节变化,进而不能用于预测植物对大气 CO_2 浓度升高的长期响应。

0.5.2 开顶式气室

开顶式气室(OTC,open top chamber)是将某一空间自然生长的植株用透明材料包围,只开放顶部与大气相通,人为控制空间内 CO_2 浓度。这种培养方式的优点是植物生长环境因子接近外界自然状态,如光照和湿度,所以可以作为长期观测平台使用,在早年受到了一些生理生态学家的信赖。例如,美国堪萨斯州立大学 Owensby 利用开顶式生长箱对堪萨斯州大草原碳储量与草种组成对高浓度 CO_2 的响应进行了为期 8 年的定位调查(Owensby et al.,1999);日本广岛大学 Wang 利用同样的方法在广岛进行连续 2 年的定位实验,研究大气 CO_2 浓度升高对青刚栎生长的影响(Wang et al.,2012)等。但是,气室内空气相对静止,温度高于外界,增加植物蒸腾,影响生长速率,给长期定位观测结果造成误差。

0.5.3 自由大气 CO_2 富集试验

自由大气 CO_2 富集试验(FACE,Free-air CO_2 enrichment)是在植株生长的小区内布设一系列管道,将 CO_2 通入小区,浓度由中心处理器控制。这种方法使植物生长环境最接近自然状态,生长空间大,适合进行研究植物生理生态响应的长期性试验,其结果具有很强的代表性。例如,约克大学 Schneider 等利用 FACE 实验平台在英国进行了连续 10 年的定位观测,研究高 CO_2 浓度对黑麦草(*Lolium perenne* L.)草地生态系统氮循环的影响(Schneider et al.,2004);丹麦科技大学 Klaus 等利用 FACE 平台对石南生态系统氮循环对高 CO_2 浓度环境响应进行研究(Larsen et al.,2011),取得了一系列丰富的数据资料,为植物对未来气候变化的长期响应研究积累了可靠资料。

大气 CO_2 浓度升高条件下的封闭式(如温室、人工气候室或是生长箱)与半开放式环境控制平台(如 OTC)可以提供重复性较高的环境控制试验。而在大田条件下,不存在两个气候条件完全相同的生长季。但是,在封闭式的人工模拟控制平台与开放式的 FACE 平台下得出的试验结论往往存在差异。近年一项 meta 分析表明,封闭式人工控制环境与 FACE 平台的光合特性与产量方面等实验结果的相关性仅为 0.26(Poorter et al.,2016)。OTC 虽然可以提供半开放式的试验条件,但 meta 分析的结果表明,其与 FACE 条件下的试验结果仍具有较大的差异(de Graaff et al.,2006)。比如,OTC 条件下作物产量的增加幅度常常大于 FACE 条件(Ainsworth,2008)。这些结果都表明了利用 FACE 试验平台预测未来气候变化下作物产量形成响应策略的重要性。另外,FACE 平台下生长的作物群体常常显著大于人工模拟控制环境平台。所以人工模拟控制环境平台适用于个体——分子尺度的研究类型,而 FACE 平台可以进行较大尺度的群体适应性研究,如土壤微生物群落适应性、多种基因型作物的适应性与产量形成规律(Bourgault et al.,2018;Köhler et al.,2019;Ainsworth et al.,2020;Hao et al.,2012;2016)。

0.6 当前研究中存在的主要问题

(1)大气 CO_2 浓度增加对陆地生态系统物质平衡与循环具有深远影响。但是,在与增温、干旱与营养元素匮乏的交互作用下,CO_2 的气肥作用具有区域多样性,且容易被完全控制条件平台下的研究结果放大或缩小。不同区域下,作物生长与产量形成对大气 CO_2 浓度升高响应的差异亟待进一步阐明,对于揭示未来作物生产与粮食安全具有重要意义。

(2)前人关于大气 CO_2 浓度升高下,C_3 植物光合能力等适应机制已进行过较多研究。而 C_4 植物由于光合作用过程中叶内 CO_2 利用效率较高,被认为在 CO_2 升高下提高潜力较小。但近年来,已有不少研究表明,C_4 植物在高 CO_2 下的适应能力也受到供氮水平、灌水条件、物质分配与利用等多重因素影响,所以其对 CO_2 浓度升高的响应过程与机制研究亟待加强。

(3)目前,在未来气候变化背景下,适用于不同区域的作物基因型与作物管理模式相结合的整体产量提升技术研究相对薄弱。亟待通过跨区域的多点试验,将作物基因型与管理模式相结合,优化增产策略,减少未来环境变化对农业生态系统造成的负担,并创新技术提高产量,在FACE平台进行检测验证,为解决农业发展困境提供有力保障。

参考文献

樊艳文,王襄平,曾令兵,等,2011.北京栓皮栎林胸径-树高相关生长关系的分析[J].北京林业大学学报,33(6):146-150.

范志强,王政权,吴楚,等,2004.不同供氮水平对水曲柳苗木生物量、氮分配及其季节变化的影响[J].应用生态学报,15(9):1497-1501.

高亚军,李生秀,2003.黄土高原地区农田水氮效应[J].植物营养与肥料学报,9(1):14-18.

韩雪,郝兴宇,王贺然,等,2012.FACE条件下冬小麦生长特征及产量构成的影响[J].中国农学通报,28(36):154-159.

林世青,卢从明,刘丽娜,1996.二氧化碳加富对大豆叶片光系统Ⅱ功能的影响[J].植物生态学报,20(6):517-523.

卢从明,张其德,1996.CO_2加富对杂交稻及其亲本原初光能转化效率和激发能分配的影响[J].生物物理学报,12(4):731-734.

平晓燕,周广胜,孙敬松,2010.植物光合产物分配及其影响因子研究进展[J].植物生态学报,34(1):100-111.

任书杰,曹明奎,陶波,等,2006.陆地生态系统氮状态对碳循环的限制作用研究进展[J].地理科学进展,25(4):58-67.

山仑,邓西平,康绍忠,2002.我国半干旱地区农业用水现状及发展方向[J].水利学报,9(9):27-31.

上官周平,李世清,2004.旱地作物氮素营养生理生态[M].北京:科学出版社.

唐如航,郭连旺,陈根云,1998.大气CO_2浓度倍增对水稻光合速率和Rubisco的影响[J].植物生理学报,24(3):309-312.

王忠强,吴良欢,刘婷婷,等,2007.供氮水平对爬山虎(*Parthenocissus*

tricuspidata Planch)生物量及养分分配的影响[J]. 生态学报,27(8):3435-3441.

许振柱,周广胜,2007. 全球变化下植物的碳氮关系及其环境调节研究进展——从分子到生态系统[J]. 植物生态学报,31(4):738-747.

薛香,吴玉娥,陈荣江,等,2006. 小麦籽粒灌浆过程的不同数学模型模拟比较[J]. 麦类作物学报,26(6):169-171.

于贵瑞,方华军,伏玉玲,等,2011. 区域尺度陆地生态系统碳收支及其循环过程研究进展[J]. 生态学报,31(19).

张其德,卢从明,刘丽娜,等,1997. CO_2浓度倍增对垂柳和杜仲叶绿体吸收光能和激发能分配的影响[J]. 植物学报,39(9):845-848.

朱艳,曹卫星,王其猛,等,2004. 基于知识模型和生长模型的小麦管理决策支持系统[J]. 中国农业科学,37(6):814-820.

Ainsworth E A, 2008. Rice production in a changing climate: A meta-analysis of responses to elevated carbon dioxide and elevated ozone concentration [J]. Global Change Biology,14:1642-1650.

Ainsworth E A, Alistair R, Leakey A D B, et al, 2007. Does elevated atmospheric [CO_2] alter diurnal C uptake and the balance of C and N metabolites in growing and fully expanded soybean leaves? [J]. Journal of Experimental Botany (3):579-591.

Ainsworth E A, Bush D R, 2010. Carbohydrate Export from the Leaf: A Highly Regulated Process and Target to Enhance Photosynthesis and Productivity [J]. Plant Physiology,155(1):64-69.

Ainsworth E A, Long S P, 2005. What have we learned from 15 years of free-air CO_2 enrichment (FACE)? A meta-analytic review of the responses of photosynthesis, canopy properties and plant production to rising CO_2[J]. New Phytologist,165(2):351-372.

Ainsworth E A, Long S P, 2020. 30 years of free-air carbon dioxide enrichment (FACE): What have we learned about future crop productivity and its potential for adaptation? [J]. Global Change Biology.

Ainsworth E A, Rogers A, 2007. The response of photosynthesis and stomatal conductance to rising [CO_2]: mechanisms and environmental interactions [J]. Plant, Cell and Environment,30(3):258-270.

Albert K R, Mikkelsen T N, Michelsen A, et al, 2011. Interactive effects of drought, elevated CO_2 and warming on photosynthetic capacity and photosystem performance in temperate heath plants [J]. J Plant Physiology, 168 (13):1550-1561.

Andrews M, Raven J A, Sprent J I, 2010. Environmental effects on dry matter partitioning between shoot and root of crop plants: relations with growth and shoot protein concentration [J]. Annals of Applied Biology, 138 (1):57-68.

Anten N P R, Schieving F, Werger M J A, 1995. Patterns of Light and Nitrogen Distribution in Relation to Whole Canopy Carbon Gain in C_3 and C_4 Monocotyledonous and Dicotyledonous Species [J]. Oecologia, 101(4):504-513.

Becker T W, Perrot-Rechenmann C, Suzuki A, et al, 1993. Subcellular and immunocytochemical localization of the enzymes involved in ammonia assimilation in mesophyll and bundle-sheath cells of maize leaves [J]. Planta, 191 (1): 129-136.

Bernacchi C J, Kimball B A, Quarles D R, et al, 2007. Decreases in stomatal conductance of soybean under open-air elevation of CO_2 are closely coupled with decreases in ecosystem evapotranspiration [J]. Plant Physiology, 143:134-144.

Bernacchi C J, VanLoocke A, 2015. Terrestrial ecosystems in a changing environment: A dominant role for water [J]. Annual Review of Plant Biology, 66:599-622.

Bindi M, Fibbi L, Miglietta F, 2001. Free Air CO_2 Enrichment (FACE) of grapevine (*Vitis vinifera* L.): II. Growth and quality of grape and wine in response to elevated CO_2 concentrations [J]. European Journal of Agronomy, 14:145-155.

Bishop K A, Betzelberger A M, Long S P, et al, 2015. Is there potential to adapt soybean (*Glycine max* Merr.) to future [CO_2]? An analysis of the yield response of 18 genotypes in free-air CO_2 enrichment [J]. Plant, Cell and Environment, 38(9):1765-1774.

Bloom A J, Burger R, Asensior S R, et al, 2010. Carbon Dioxide Enrichment Inhibits Nitrate Assimilation in Wheat and Arabidopsis [J]. Science, 328 (5980):899-903.

Bobich E, Barron-Gaford G, Rascher K, et al, 2010. Effects of drought and changes in vapour pressure deficit on water relations of Populus deltoides growing in ambient and elevated CO_2 [J]. Tree Physiology, 30(7):866-875.

Bourgault M, Brand J, Tausz M, et al, 2016. Yield, growth and grain nitrogen response to elevated CO_2 of five field pea (*Pisum sativum* L.) cultivars in a low rainfall environment [J]. Field Crop Research. 196:1-9.

Bourgault M, Löw M, Tausz-Posch S, et al, 2018. Effects of a heat wave on lentil growth under free-air CO_2 enrichment (FACE) in a semi-arid environment [J]. Crop Science. 58:803-812.

Cai C, Yin X, He S, et al, 2016. Responses of wheat and rice to factorial combinations of ambient and elevated CO_2 and temperature in FACE experiments [J]. Global Change Biology, 22:856-874.

Casson S, Gray J E, 2008. Influence of environmental factors on stomatal development. New Phytologist [J]. 178 (1):9-23.

Chapin Ⅲ F S, Schulze E D, Mooney H A, 1990. The ecology and economics of storage in plants [J]. Annual Review of Ecology and Systematics, 21:423-447.

Cheng D L, Niklas K J, 2007. Above-and below-ground biomass relationships across 1534 forested communities [J]. Annals of Botany, 99 (1):95-102.

Cheng L, Zhu J G, Chen G, et al, 2010. Atmospheric CO_2 enrichment facilitates cation release from soil [J]. Ecological Letters, 13:284-291.

Cheng L, Fuchigami L H, 2002. Growth of young apple trees in relation to reserve nitrogen and carbohydrates [J]. Tree Physiology, 22(18):1297-1303.

Cheng W, Sakai H, Yagi K, et al, 2010. Combined effects of elevated [CO_2] and high night temperature on carbon assimilation, nitrogen absorption, and the allocations of C and N by rice (*Oryza sativa* L.) [J]. Agricultural and Forest Meteorology, 150 (9):1174-1181.

Cotrufo M, Gorissen A, 1997. Elevated CO_2 enhances below-ground C allocation in three perennial grass species at different levels of N availability [J]. New Phytologist, 137(3):421-431.

de Graaff M A, VanGroenigen K J, Six J, et al, 2006. Interactions between plant growth and soil nutrient cycling under elevated CO_2: A meta-analysis [J]. Global Change Biology, 12:2077-2091.

Dewar R, 1993. A root-shoot partitioning model based on carbon-nitrogen-water interactions and Munch phloem flow [J]. Functional Ecology, 7(3):356-368.

Dier M, Hüther L, Schulze W X, et al, 2020. Elevated atmospheric CO_2

concentration has limited effect on wheat grain quality regardless of nitrogen supply [J]. Journal of Agricultural and Food Chemistry,68:3711-3721.

Dier M,Meinen R,Erbs M,et al,2018. Effects of free air carbon dioxide enrichment (FACE) on nitrogen assimilation and growth of winter wheat under nitrate and ammonium fertilization [J]. Global Change Biology,24:40-54.

Dong Y L,Wang Z Y,Sun H,et al,2018. The response patterns of arbuscular mycorrhizal and ectomycorrhizal symbionts under elevated CO_2: A meta-analysis [J]. Frontiers in Microbiology,9:1248.

Drake B G, GonzalezMeler M A, Long S P, 1997. More efficient plants: A consequence of rising atmospheric CO_2? [J]. Annual Review of Plant Physiology and Plant Molecular Biology,48:609-639.

Dreccer M F,Van Oijen M,Schapendonk A H C M,et al,2000. Dynamics of vertical leaf nitrogen distribution in a vegetative wheat canopy. Impact on canopy photosynthesis [J]. Annals of Botany,86 (4):821-831.

Dukes J S,Chiariello N R,Cleland E E,et al,2005. Responses of grassland production to single and multiple global environmental changes [J]. PLoS Biology,3(10): 1829-1837.

Duursma R, Barton C, Eamus D, et al, 2011. Rooting depth explains [CO_2] × drought'interaction in *Eucalyptus saligna* [J]. Tree Physiology,31:922-931.

Dyckmans J,Flessa H,2001. Influence of tree internal N status on uptake and translocation of C and N in beech: a dual ^{13}C and ^{15}N labeling approach [J]. Tree Physiology,21(6):395-401.

Dyckmans J, Flessa H, 2005. Partitioning of remobilised N in young beech (*Fagus sylvatica* L.)is not affected by elevated [CO_2] [J]. Annals of Forest Science,62 (3):285-288.

Dyckmans J, Flessa H, Polle A, et al, 2000. The effect of elevated [CO_2] on uptake and allocation of ^{13}C and ^{15}N in beech (*Fagus sylvatica* L.) during leafing [J]. Plant Biology,2 (1):113-120.

Ellsworth D S, Reich P B, Naumburg E S, et al, 2004. Photosynthesis, carboxylation and leaf nitrogen responses of 16 species to elevated pCO_2 across four free-air CO_2 enrichment experiments in forest, grassland and desert [J]. Global Change Biology,10(12):2121-2138.

Erbs M, Manderscheid R, Huther L, 2015. Free-air CO_2 enrichment modifies

第一部分 大气 CO_2 浓度升高对作物生长影响的研究进展

maize quality only under drought stress [J]. Agronomy for Sustainable Development,35:203-212.

Farrar J,Williams M,2006. The effects of increased atmospheric carbon dioxide and temperature on carbon partitioning,source-sink relations and respiration [J]. Plant, Cell and Environment,14(8):819-830.

Farrar J F,Gunn S,1996. Effects of temperature and atmospheric carbon dioxide on source-sink relations in the context of climate change. In: Zamski E, Schaffer AA (eds) Photoassimilate Distribution and Crops. Source-Sink Relationships [J]. Marcel Dukker Inc. ,New York,389-406.

Feldman D R,Collins W D,Gero P J,et al,2015. Observational determination of surface radiative forcing by CO_2 from 2000 to 2010 [J]. Nature,519 (7543): 339-343.

Field C,Mooney H,1986. The photosynthesis-nitrogen relationship in wild plants. In: Givnish T (ed) On the Economy of Plant form and Function [M]. Cambridge University Press,Cambridge,UK,25-55.

Fitzgerald G J, Tausz M, Leary G, et al, 2016. Elevated atmospheric [CO_2] can dramatically increase wheat yields in semi-arid environments and buffer against heat waves [J]. Global Change Biology,22:2269-2284.

Furbank RT,Chitty JA,von Caemmerer S,et al,1996. Antisense RNA inhibition of *RbcS* gene expression reduces rubisco level and photosynthesis in the C_4 plant *Flaveria bidentis* [J]. Plant Physiology,111 (3):725-734.

Gao J, Han X, Seneweera S, et al, 2015. Leaf photosynthesis and yield components of mung bean under fully open-air elevated [CO_2] [J]. Journal of Integrative Agriculture,20 (5):977-983.

Gastal F,Lemaire G,2002. N uptake and distribution in crops:an agronomical and ecophysiological perspective [J]. Journal of Experimental Botany, 53 (370):789-799.

Gastal F, Saugier B, 2006. Relationships between nitrogen uptake and carbon assimilation in whole plants of tall fescue [J]. Plant,Cell and Environment,12 (4):407-418.

Ge Z M, Zhou X, Kellomäki S, 2012. Carbon assimilation and allocation (^{13}C labeling) in a boreal perennial grass(*Phalaris arundinacea*)subjected to elevated temperature and CO_2 through a growing season [J]. Environmental and

Experimental Botany,75:150-158.

Ghannoum O,Von Caemmerer S,Ziska L H,et al,2000a. The growth response of C4 plants to rising atmospheric CO_2 partial pressure: a reassessment [J]. Plant, Cell and Environment,23(9):931-942.

Ghannoum O, Evans J R, Caemmerer S, 2011. Chapter 8 Nitrogen and Water Use Efficiency of C4 Plants [R]. In: C4 Photosynthesis and Related CO_2 Concentrating Mechanisms. 129-146.

Ghini R, Torre-Neto A, Dentzien A F M, et al, 2015. Coffee growth, pest and yield responses to free-air CO_2 enrichment [J]. Climatic Change. 132: 307-320.

Gray S B,Dermody O,Klein S P,et al,2016. Intensifying drought eliminates the expected benefits of elevated carbon dioxide for soybean [J]. Nature Plants,2: 16132.

Gray S B, Strellner R S, Puthuval K K, , 2013. Minirhizotron imaging reveals that nodulation of field-grown soybean is enhanced by free-air CO_2 enrichment only when combined with drought stress [J]. Functional Plant Biology, 40: 137-147.

Gutierrez D,Gutierrez E,Perez P,et al,2009. Acclimation to future atmospheric CO_2 levels increases photochemical efficiency and mitigates photochemistry inhibition by warm temperatures in wheat under field chambers [J]. Physiology Plantarum,137(1):86-100.

Högy P,Brunnbauer M,Koehler P,2013. Grain quality characteristics of spring wheat (*Triticum aestivum*) as affected by free-air CO_2 enrichment [J]. Environmental and Experimental Botany,88:11-18.

Högy P,Keck M,Niehaus K,et al,2010. Effects of atmospheric CO_2 enrichment on biomass, yield and low molecular weight metabolites in wheat grain [J]. Journal of Cereal Science,52:215-220.

Högy P,Wieser H,Köhler P,et al,2009. Effects of elevated CO_2 on grain yield and quality of wheat: Results from a 3-year free-air CO_2 enrichment experiment [J]. Plant Biology,11:60-69.

Hao X Y, Han X, Lam S K, , 2012. Effects of fully open-air [CO_2] elevation on leaf ultrastructure, photosynthesis, and yield of two soybean cultivars [J]. Photosynthetica,50 (3):362-370.

Hao X Y,Li P,Han X,et al,2016. Effects of free-air CO_2 enrichment (FACE) on N, P and K uptake of soybean in northern China [J]. Agricultural and Forest Meteorology,218:216-266.

Hasegawa T,Sakai H,Tokida T,et al,2013. Rice cultivar responses to elevated CO_2 at two free-air CO_2 enrichment (FACE) sites in Japan [J]. Functional Plant Biology,40:148-159.

Hasegawa T,Sakai H,Tokida T,et al,2019. A high-yielding rice cultivar "Takanari" shows no N constraints on CO_2 fertilization [J]. Frontiers in Plant Science,10:15.

Hessen D O, Agren G I, Anderson T R, et al, 2004. Carbon, sequestration in ecosystems:The role of stoichiometry [J]. Ecology,85 (5):1179-1192.

Hikosaka K,Terashima I,Katoh S,1994. Effects of leaf age, nitrogen nutrition and photon flux-density on the distribution of nitrogen among leaves of a Vine (*Ipomoea-Tricolor* Cav.) grown horizontally to avoid mutual shading of leaves [J]. Oecologia,97 (4):451-457.

Hoch G,Korner C,2003. The carbon charging of pines at the climatictreeline:a global comparison [J]. Oecologia,135 (1):10-21.

Hoch G, Richter A, Körner C, 2003. Non-structural carbon compounds in temperate forest trees [J]. Plant,Cell and Environment,26 (7):1067-1081.

Houghton R A, 2001. Counting terrestrial sources and sinks of carbon [J]. Climatic Change,48 (4):525-534.

Houshmandfar A, Fitzgerald G J, Macabuhay A A, et al, 2016. Trade-offs between water-use related traits, yield components and mineral nutrition of wheat under free-air CO_2 enrichment (FACE) [J]. European Journal of Agronomy,76:66-74.

Houshmandfar A, Fitzgerald G J, O'Leary G, et al, 2018. The relationship between transpiration and nutrient uptake in wheat changes under elevated atmospheric CO_2[J]. Physiologia Plantarum,163:516-529.

Hussain M Z, VanLoocke A, Siebers M H, et al, 2013. Future carbon dioxide concentration decreases canopy evapotranspiration and soil water depletion by field-grown maize [J]. Global Change Biology,19:1572-1584.

IPCC,2013. Summary for Policymakers. Climate Change:The Physical Science Basis. Contribution of Working Group I to the Fifth Assessment Report of the Intergovernmental Panel on Climate Change [J]. Cambridge University Press,

Cambridge: United Kingdom and New York, NY, USA.

Iversen C M, Ledford J, Norby R J, 2008. CO_2 enrichment increases carbon and nitrogen input from fine roots in a deciduous forest [J]. New Phytologist, 179 (3): 837-847.

Jackson R B, Mooney H A, Schulze E D, 1997. A global budget for fine root biomass, surface area, and nutrient contents [J]. Proceedings of the National Academy of Sciences of the United States of America, 94 (14): 7362-7366.

Johnson S L, Lincoln D E, 2001. Allocation responses to CO_2 enrichment and defoliation by a native annual plant *Heterotheca subaxillaris* [J]. Global Change Biology, 6 (7): 767-778.

Köhler I H, Huber S C, Bernacchi C J, et al, 2019. Increased temperatures may safeguard the nutritional quality of crops under future elevated CO_2 concentrations [J]. Plant Journal, 97: 872-886.

Keymer A, Pimprikar P, Wewer V, et al, 2017. Lipid transfer from plants to arbuscular mycorrhiza fungi [J]. Elife. 6: e29107.

Kimball B A, 2016. Crop responses to elevated CO_2 and interactions with H_2O, N, and temperature [J]. Current Opinion in Plant Biology, 31: 36-43.

Larsen K S, Andresen L C, Beier C, et al, 2011. Reduced N cycling in response to elevated CO_2, warming, and drought in a Danish heathland: Synthesizing results of the CLIMAITE project after two years of treatments [J]. Global Change Biology, 17 (5): 1884-1899.

Leakey A D B, 2009. Rising atmospheric carbon dioxide concentration and the future of C_4 crops for food and fuel. Proceedings of the Royal Society B [J]. Biological Sciences, 276: 1666.

Leakey A D, Bernacchi C J, Ort D R, et al, 2010. Long-term growth of soybean at elevated [CO_2] does not cause acclimation of stomatal conductance under fully open-air conditions [J]. Plant, Cell and Environment, 29 (9): 1794-1800.

Leakey A D B, Ainsworth E A, Bernacchi C J, et al, 2009. Elevated CO_2 effects on plant carbon, nitrogen, and water relations: six important lessons from FACE [J]. Journal of Experimental Botany, 60 (10): 2859-2876.

Leakey A D B, Uribelarrea M, Ainsworth E A, et al, 2006. Photosynthesis, productivity, and yield of maize are not affected by open-air elevation of CO_2 concentration in the absence of drought. Plant Physiology. 140 (2): 779-790.

Liu S R, Barton C, Lee H, et al, 2002. Long-term response of Sitka spruce (*Picea sitchensis* (Bong.) Carr.) to CO_2 enrichment and nitrogen supply. I. Growth, biomass allocation and physiology [J]. Plant Biosystems-An International Journal Dealing with all Aspects of Plant Biology, 136 (2):189-198.

Loladze I, 2002. Rising atmospheric CO_2 and human nutrition: Toward globally imbalanced plant stoichiometry? [J]. Trends in Ecology and Evolution, 17:457-461.

Loladze I, 2014. Hidden shift of the ionome of plants exposed to elevated CO_2 depletes minerals at the base of human nutrition. Elife. 3:e02245.

Long S P, Ainsworth E A, Leakey A D B, et al, 2006. Food for thought: Lower-than-expected crop yield stimulation with rising CO_2 concentrations [J]. Science, 312 (5782):1918-1921.

Long S P, Ainsworth E A, Rogers A, et al, 2004. Rising atmospheric carbon dioxide: Plants face the future [J]. Annual Review of Plant Biology, 55: 591-628.

Luo Y, Su B, Currie W S, et al, 2004. Progressive nitrogen limitation of ecosystem responses to rising atmospheric carbon dioxide [J]. Bioscience, 54 (8):731-739.

Macabuhay A, Houshmandfar A, Nuttall J, et al, 2018. Can elevated CO_2 buffer the effects of heat waves on wheat in a dryland cropping system? [J]. Environmental and Experimental Botany. 155:578-588.

Magliulo V, Bindi M, Rana G, 2003. Water use of irrigated potato (*Solanum tuberosum* L.) grown under free air carbon dioxide enrichment in central Italy [J]. Agriculture Ecosystems & Environment, 97:65-80.

Manderscheid R, Dier M, Erbs M, et al, 2018. Nitrogen supply-A determinant in water use efficiency of winter wheat grown under free air CO_2 enrichment [J]. Agricultural Water Management (210):70-77.

Manderscheid R, Erbs M, Burkart S, et al, 2016. Effects of free-air carbon dioxide enrichment on sap flow and canopy microclimate of maize grown under different water supply [J]. Journal of Agronomy and Crop Science, 202: 255-268.

Manderscheid R, Erbs M, Weigel H J, 2014. Interactive effects of free-air CO_2 enrichment and drought stress on maize growth [J]. European Journal of

Agronomy,52:11-21.

Manter D K,Kavanagh K L,Rose C L,2005. Growth response of Douglas-fir seedlings to nitrogen fertilization:importance of Rubisco activation state and respiration rates [J]. Tree Physiology,25 (8):1015-1021.

Martin C, Glover B J, 2007. Functional aspects of cell patterning in aerial epidermis [J]. Current Opinion in Plant Biology,10 (1):70-82.

Martinez-Carrasco R, Perez P, Morcuende R, 2005. Interactive effects of elevated CO_2, temperature and nitrogen on photosynthesis of wheat grown under temperature gradient tunnels [J]. Environmental and Experimental Botany,54 (1): 49-59.

Mauney J R,Kimball B A,Pinter P J J,et al,1994. Growth and yield of cotton in response to a free-air carbon dioxide enrichment (FACE) environment [J]. Agricultural and Forest Meteorology,70:49-67.

McCarthy M,Enquist B,2007. Consistency between an allometric approach and optimal partitioning theory in global patterns of plant biomass allocation [J]. Functional Ecology,21 (4):713-720.

McGrath J M,Lobell D B,2013. Reduction of transpiration and altered nutrient allocation contribute to nutrient decline of crops grown in elevated CO_2 concentrations [J]. Plant,Cell and Environment,36 (3):697-705.

Meng F, Zhang J, Yao F, et al, 2014. Interactive effects of elevated CO_2 concentration and irrigation on photosynthetic parameters and yield of maize in northeast China. PLoS One. 9:98318.

Miglietta F,Magliulo V,Bindi M,et al,1998. Free air CO_2 enrichment of potato (*Solanum tuberosum* L.):Development,growth and yield [J]. Global Change Biology (4):163-172.

Millard P,Grelet G,2010. Nitrogen storage and remobilization by trees:ecophysiological relevance in a changing world [J]. Tree Physiology,30 (9):1083-1095.

Millard P,Hester A,Wendler R,et al,2001. Interspecific defoliation responses of trees depend on sites of winter nitrogen storage [J]. Functional Ecology,15 (4):535-543.

Millard P,Proe M F,1992. Storage and internal cycling of nitrogen in relation to seasonal growth of sitka spruce [J]. Tree Physiology,10 (1):33-43.

Millard P, Sommerkorn M, Grelet GA, 2007. Environmental change and carbon limitation in trees: a biochemical, ecophysiological and ecosystem appraisal [J]. New Phytologist,175,(1):11-28.

Millard P, Thomson C,1989. The effect of the autumn senescence of leaves on the internal cycling of nitrogen for the spring growth of apple trees [J]. Journal of Experimental Botany,40 (11):1285-1289.

Milroy S P, Bange M P, Sadras V O,2001. Profiles of leaf nitrogen and light in reproductive canopies of cotton (*Gossypium hirsutum*) [J]. Annals of Botany, 87 (3):325-333.

Mooney H, Gulmon S,1979. Environmental and evolutionary constraints on the photosynthetic characteristics of higher plants, vol 316 [M]. Columbia University Press, New York.

Myers S S, Zanobetti A, Kloog I P et al,2014. Increasing CO_2 threatens human nutrition [J]. Nature,510 (7503):139-142.

Nambiar E, Fife D,1987. Growth and nutrientretranslocation in needles of radiata pine in relation to nitrogen supply [J]. Annals of Botany,60 (2):147-156.

Niklas K J, Enquist B J, 2002. Canonical rules for plant organ biomass partitioning and annual allocation [J]. American Journal of Botany,89 (5):812-819.

Nowak R S, Zitzer S F, Babcock D, et al,2004. Elevated atmospheric CO_2 does not conserve soil water in the Mojave Desert [J]. Ecology,85 (1):93-99.

Ort D R, Ainsworth E A, Aldea M, et al,2006. SoyFACE: The effects and interactions of elevated CO_2 and O_3 on soybean [C]. In: Nosberger J, Long S P, Norby R J, Stitt M, Hendrey G R, Blum H(eds)Managed ecosystems and CO_2:Case studies processes and perspectives. 71-86.

Ottman M J, Kimball B A, Pinter P J, et al,2001. Elevated CO_2 increases sorghum biomass under drought conditions [J]. New Phytologist, 150: 261-273.

Owensby C E, Ham J, Knapp A, et al,1999. Biomass production and species composition change in a tallgrass prairie ecosystem after long-term exposure to elevated atmospheric CO_2[J]. Global Change Biology,5 (5):497-506.

Pa N,2001. Carbon allocation in calcareous grassland under elevated CO_2: a combined ^{13}C pulse-labelling/soil physical fractionation study [J]. Functional Ecology,15 (1):43-50.

Palacio S, Hester A, Maestro M, et al,2008. Browsed Betula pubescent trees are not carbon-limited [J]. Functional Ecology,22 (5):808-815.

Parvin S, Uddin S, Bourgault M, et al, 2018. Water availability moderates N_2 fixation benefit from elevated [CO_2]:A 2-year free-air CO_2 enrichment study on lentil (*Lens culinaris* MEDIK.) in a water limited agroecosystem [J]. Plant,Cell and Environment,41:2418-2434.

Parvin S, Uddin S, Tausz-Posch S, et al, 2019. Elevated CO_2 improves yield and N_2 fixation but not grain N concentration of faba bean (*Vicia faba* L.) subjected to terminal drought [J]. Environmental and Experimental Botany, 165:161-173.

Poorter H, 1993. Interspecific variation in the growth response of plants to an elevated ambient CO_2 concentration [J]. Plant Ecology,104 (1):77-97.

Poorter H, Fiorani F, Pieruschka R, et al, 2016. Pampered inside, pestered outside? Differences and similarities between plants growing in controlled conditions and in the field [J]. New Phytologist,212:838-855.

Poorter H, Nagel O, 2000. The role of biomass allocation in the growth response of plants to different levels of light, CO_2, nutrients and water: a quantitative review [J]. Functional Plant Biology,27 (12):1191-1197.

Reich P B, Hobbie S E, Lee T, et al, 2006a. Nitrogen limitation constrains sustainability of ecosystem response to CO_2 [J]. Nature,440 (7086):922-925.

Reich P B, Hungate B A, Luo Y, 2006b. Carbon-nitrogen interactions in terrestrial ecosystems in response to rising atmospheric carbon dioxide [J]. Annual Review of Ecology, Evolution, and Systematics,37:611-636.

Reich P B, Knops J, Tilman D, et al, 2001. Plant diversity enhances ecosystem responses to elevated CO_2 and nitrogen deposition [J]. Nature,410 (6830): 809-812.

Robinson J M, 1988. Spinach leaf chloroplast CO_2 and NO_2 photoassimilations do not compete for photogenerated reductant manipulation of reductant levels by quantum flux density titrations [J]. Plant physiology,88 (4):1373-1380.

Ruehr N K, Offermann C A, Gessler A, et al, 2009. Drought effects on allocation of recent carbon: from beech leaves to soil CO_2 efflux [J]. New Phytologist, 184 (4):950-961.

Sage R F, Kubien D S, 2003. Quo vadis C_4? An ecophysiological perspective on global change and the future of C_4 plants [J]. Photosynth Research,77 (2-3):

209-225.

Sakai H, Tokida T, Usui Y, et al, 2019. Yield responses to elevated CO_2 concentration among Japanese rice cultivars released since 1882 [J]. Plant Production Science,22:352-366.

Sarker B C, Hara M, 2011. Effects of elevated CO_2 and water stress on the adaptation of stomata and gas exchange in leaves of eggplants (*Solanum melongena* L.) [J]. Bangladesh Journal of Botany,40 (1):1-8.

Sauter J J, Vancleve B,1994. Storage, Mobilization and Interrelations of Starch, Sugars, Protein and Fat in the Ray Storage Tissue of Poplar Trees [J]. Trees-Struct Funct. 8 (6):297-304.

Schneider M K, Lüscher A, Richter M, et al, 2004. Ten years of free-air CO_2 enrichment altered the mobilization of N from soil in *Lolium perenne* L. swards [J]. Global Change Biology. 10 (8):1377-1388.

Sekhar K M, Kota V R, Reddy T P, et al, 2020. Amelioration of plant responses to drought under elevated CO_2 by rejuvenating photosynthesis and nitrogen use efficiency: implications for future climate-resilient crops [J]. Photosynthesis Research.

Sheen J,1994. Feedback-Control of Gene Expression. Photosynthesis Research [J]. 39 (3):427-438.

Shimono H, Okada M, Yamakawa Y, et al,2009. Genotypic variation in rice yield enhancement by elevated CO_2 relates to growth before heading, and not to maturity group [J]. Journal of Experimental Botany,60:523-532.

Shiraiwa T, Sinclair T R, 1993. Distribution of Nitrogen among Leaves in Soybean Canopies [J]. Crop Science,33 (4):804-808.

Silla F, Escudero A, 2003. Uptake, demand and internal cycling of nitrogen in saplings of Mediterranean Quercus species [J]. Oecologia,136 (1):28-36.

Singh S D, Chakrabarti B, Muralikrishna K S, et al,2013. Yield response of important field crops to elevated air temperature and CO_2 level [J]. Indian Journal of Agricultural Sciences,83:1009-1012.

Smith N G, Keenan T F, 2020. Mechanisms underlying leaf photosynthetic acclimation to warming and elevated CO_2 as in-ferred from least-cost optimality theory [J]. Global Change Biology,26:5202-5216.

Smith S E, Smith F A, 2012. Fresh perspectives on the roles of arbuscular mycorrhizal fungi in plant nutrition and growth [J]. Mycologia,104:1-13.

Spann T M, Beede R H, Dejong T M, 2008. Seasonal carbohydrate storage and mobilization in bearing and non-bearing pistachio (*Pistacia vera*) trees [J]. Tree Physiology,28 (2):207-213.

Sreeharsha R, Mudalkar S, Sengupta D, et al, 2019. Mitigation of drought-induced oxidative damage by enhanced carbon assimilation and an efcient antioxidative metabolism under high CO_2 environment in pigeonpea (*Cajanus cajan* L.)[J]. Photosynthesis Research,139:425-439.

Stitt M,1991. Rising CO_2 levels and their potential significance for carbon flow in photosynthetic cells [J]. Plant, Cell & Environment,14 (8):741-762.

Stocker T, 2013. Technical Summary. In: Climate Change 2013: The Physical Science Basis. Contribution of Working Group I to the Fifth Assessment Report of the Intergovernmental Panel on Climate Change [J]. Computational Geometry,18 (2):95-123.

Strullu-Derrien C, Selosse M A, Kenrick P, et al, 2018. The origin and evolution of mycorrhizal symbioses: From palaeomycology to phylogenomics [J]. New Phytologist,220:1012-1030.

Suter D, Frehner M, Fischer B U, et al, 2002. Elevated CO_2 increases carbon allocation to the roots of *Lolium perenne* under free-air CO_2 enrichment but not in a controlled environment [J]. New Phytologist,154 (1):65-75.

Taiz L, Zeiger E, 2006. Plant Physiology. Fourthedn. Sinauer Associates, Inc., Sunderland.

Taub D R, Miller B, Allen H, 2008. Effects of elevated CO_2 on the protein concentration of food crops: A meta-analysis [J]. Global Change Biology,14: 565-575.

Taub D R, Wang X Z, 2008. Why are nitrogen concentrations in plant tissues lower under elevated CO_2? A critical examination of the hypotheses, Journal of Integrative Plant Biology,50:1365-1374.

Tausz-Posch S, Dempsey R W, Seneweera S, et al, 2015. Does a freely tillering wheat cultivar benefit more from elevated CO_2 than a restricted tillering cultivar in a water-limited environment? [J]. European Journal of Agronomy. 64:21-28.

Tilman D, 1989. Plant strategies and the dynamics and structure of plant communities [J]. Bulletin of Mathematical Biology,4 (1):28-29.

Tingey D T, Mckane R B, Olszyk D M, et al, 2003. Elevated CO_2 and temperature alter nitrogen allocation in Douglas-fir [J]. Global Change Biology,9 (7):1038-1050.

Uddling J, Broberg M C, Feng Z, et al, 2018. Crop quality under rising atmospheric CO_2[J]. Current Opinion in Plant Biology,45:262-267.

Urabe J, Elser J J, Kyle M, et al, 2002. Herbivorous animals can mitigate unfavourable ratios of energy and material supplies by enhancing nutrient recycling [J]. Ecological Letters,5 (2):177-185.

Würth M K R, Pelaez-Riedl S, Wright S J, et al, 2005. Non-structural carbohydrate pools in a tropical forest [J]. Oecologia,143 (1):11-24.

Wand S J E,Midgley G F,Jones M H,et al,1999. Responses of wild C_4 and C_3 grass (Poaceae) species to elevated atmospheric CO_2 concentration: a meta-analytic test of current theories and perceptions [J]. Global Change Biology,5 (6):723-741.

Wang W L,Cai C,He J,et al,2020. Yield,dry matter distribution and photosynthetic characteristics of rice under elevated CO_2 and increased temperature conditions [J]. Field Crop Research,248:107605.

Wang X, Nakatsubo T, Nakane K, 2012. Impacts of elevated CO_2 and temperature on soil respiration in warm temperate evergreen Quercus glauca stands:an open-top chamber experiment [J]. Ecological Research,27 (3): 595-602.

Wardlaw I, 1969. The effect of water stress on translocation in relation to photosynthesis and growth [J]. Australian journal of biological sciences, 22 (1):1-16.

Warren J, Norby R, Wullschleger S, 2011. Elevated CO_2 enhances leaf senescence during extreme drought in a temperate forest [J]. Tree Physiology,31:117-130.

Weigel H J, Manderscheid R, 2012. Crop growth responses to free air CO_2 enrichment and nitrogen fertilization:Rotating barley,ryegrass,sugar beet and wheat [J]. European Journal of Agronomy,43:97-107.

Wenxuan H,Jingyu F,2003. Allometry and its application in ecological scaling [J]. Acta Scientiarum Naturalium Universitatis Pekinensis,39 (4):583-593.

Westoby M, Falster D S, Moles A T, et al, 2002. Plant ecological strategies:

Some leading dimensions of variation between species [J]. Annual Review of Ecology and Systematics, 33: 125-159.

Wong S, 1979. Elevated atmospheric partial pressure of CO_2 and plant growth. I. Interactions of nitrogen nutrition and photosynthetic capacity in C_3 and C_4 plants [J]. Oecologia, 44: 68-74.

Wright I J, Reich P B, Westoby M, et al, 2004. The worldwide leaf economics spectrum [J]. Nature, 428 (6985): 821-827.

Xu Z, Zhou G, Wang Y, 2007. Combined effects of elevated CO_2 and soil drought on carbon and nitrogen allocation of the desert shrub Caragana intermedia [J]. Plant and Soil, 301 (1): 87-97.

Yang L X, Huang J, Yang H, et al, 2006. The impact of free-air CO_2 enrichment (FACE) and N supply on yield formation of rice crops with large panicle [J]. Field Crop Research, 98: 141-150.

Zak D R, Pregitzer K S, Curtis P S, et al, 2000. Atmospheric CO_2, soil-N availability, and allocation of biomass and nitrogen by Populus tremuloides [J]. Ecological Applications, 10 (1): 34-46.

Zhou Z C, Shangguan Z P, 2009. Effects of Elevated CO_2 Concentration on the Biomasses and Nitrogen Concentrations in the Organs of Sainfoin (*Onobrychis viciaefolia* Scop.) [J]. Agricultural Sciences in China, 8 (4): 424-430.

Zhu C, Kobayashi K, Loladze I, et al, 2018. Carbon dioxide (CO_2) levels this century will alter the protein, micronutrients, and vitamin content on rice grains with potential health consequences for the poorest rice-dependent countries [J]. Science Advances, 4: eaaq1012.

Zhu X G, Long S P, 2009. Can increase in Rubisco specificity increase carbon gain by whole canopy? A modeling analysis [R]. In: A. Laisk, Nedbal L, Govindjee (eds) Photosynthesis in silico: Understanding complexity from molecules to ecosystems. 401-416.

Zinta G, Abdelgawad H, Domagalska M A, et al, 2014. Physiological, biochemical, and genome-wide transcriptional analysis reveals that elevated CO_2 mitigates the impact of combined heat wave and drought stress in Arabidopsis thaliana at multiple organizational levels [J]. Global Change Biology, 20: 3670-3685.

第二部分　大气 CO_2 浓度升高对主要 C_3 豆科作物的影响

大豆（*Glycine max*（L.）Merr.）是世界上重要的豆科牧草,是人类、牲畜的蛋白质来源之一。中国是世界上最大的大豆进口国,进口数量逐年递增,而产量却难以提高。

在大气 CO_2 浓度升高背景下,多数作物在初期光合同化物积累速度加快,体内原有的碳氮平衡被打破,导致进一步生长受氮素限制的可能性增加。由于豆科作物根瘤菌具有固氮功能,所以有研究认为,其生长发育在 CO_2 浓度升高环境下受氮素限制的可能性较小。一项针对美国 FACE 平台 27 年试验结果的 meta 分析表明, C_3 豆科作物大豆、桃树、花生与芸豆产量在 CO_2 浓度升高下提高了约 16%（Kimball et al.,2016）。同时,大豆籽粒蛋白质含量在增产条件下并没有降低,根瘤质量增加引起的固氮量提高是主要原因之一（Ort et al.,2006）。

然而,作物对未来气候变化的响应存在较大地域性差异。在我国北方,大豆等豆科作物生长发育在大气 CO_2 浓度升高下的响应特征还有待明确,特别是光合特性、元素吸收分配、产量与品质形成,以及其与根瘤菌生长的关系。2007 年,中国北方第一个 FACE 试验平台在北京昌平建立,为探讨大气 CO_2 浓度升高背景下中国北方大田条件下豆科作物生长发育适应调节规律提供了良好的机会。这对研究我国大豆未来生产能力,提高大豆产量,保证大豆供给和指导大豆进出口有重要意义。

第1章 大气 CO_2 浓度升高对大豆元素吸收利用的影响

1.1 引言

研究表明,大气 CO_2 浓度升高会改变植株原有的养分平衡(Pang et al.,2006)、循环与利用(Zeng et al.,2011)、生长与发育状况(Hao et al.,2014),以及残体分解过程(Lam et al.,2013)。氮、磷、钾是大豆生长与发育所需的必需营养,因而了解其在大气 CO_2 浓度升高条件下的吸收利用规律至关重要。

前人对不同作物氮、磷、钾元素吸收利用在人工模拟大气 CO_2 浓度升高下适应规律的研究结果不同(Fu Sheng et al.,2002;Zeng et al.,2011;Li et al.,2015)。Ainsworth(2007)的研究表明,大气 CO_2 浓度升高提高了大豆未完全展开叶中氨基酸含量,叶片中氮同化得到增强。另一方面,CO_2 浓度升高降低了大豆生育前期的叶片氮浓度,但对中期没有显著影响,这可以归因于大豆中期植株氮需求量与自身氮固定、吸收能力的增强(Rogers et al.,2010)。水稻植株根系的钾浓度与吸收量在大气 CO_2 浓度升高下显著增强(Yang et al.,2007a)。大气 CO_2 浓度升高下,磷与钾浓度在水稻与稗草体内显著提高(Zeng et al.,2011),但在巨盘木(Kanowski,2010)体内降低,在麦珠子(Kanowski,2010)与绿豆(Li et al.,2015)体内没有明显变化。然而,CO_2 浓度升高对大田作物组织中氮、磷、钾的吸收鲜有报道(Kanowski,2010;Zeng et al.,2011;Li et al.,2015)。大豆是重要的豆科作物,大气 CO_2 浓度升高下氮、磷、钾浓度与吸收变化规律亟待研究。本研究可以为探讨 FACE 条件下,中国北方高产大豆植株氮、磷、钾的吸收与利用策略提供重要证据。

1.2 试验设计与方法

本研究于 2013—2014 年在位于北京市昌平区南部(40.13°N,116.14°E)的中国农业科学院昌平实验基地进行。田间开放式 CO_2 浓度控制通过中国农业科学院农业环境与可持续发展研究所研制的 Mini FACE 试验系统进行。Mini FACE 系统有 6 个 FACE 试验圈和 6 个对照圈,圈为正八边形,直径 4m。FACE 圈与对照圈之间的距离大于 14m,以保证 FACE 圈和对照圈中的 CO_2 浓度没有互相干扰。FACE 圈和对照圈中的 CO_2 浓度有 CO_2 传感器(Vaisala, Finland)实时监控,以保证圈内浓度达到要求。从大豆出苗到收获全生育期间,FACE 圈白天释放 CO_2,夜间不释放。CO_2 处理分别为 550 ± 60 $\mu mol/mol$(FACE)和 393 ± 40 $\mu mol/mol$(正常大气 CO_2 水平)2 个水平。FACE 圈与对照之间其他田间管理措施一致。土壤类型属褐潮土,土壤全氮 1.08 g/kg,有机质 18.27g/kg,速效钾 140.28 g/kg,碱解氮 94.83 g/kg,pH 值 8.36。试验材料为大豆中黄 35。

于大豆生长的关键生育期测定植株不同器官氮(凯氏定氮法)、磷(钼锑抗分光光度法)、钾(火焰光度法)含量与酰脲(紫外分光光度计)含量,于成熟期测定产量。

1.3 结果与分析

1.3.1 生物量

两年平均数据表明,大气 CO_2 浓度升高下大豆地上部生物量在 R1、R3、R5 和成熟期分别显著增加了 19%、30%、35%和 17%,根系生物量分别显著增加了 16%、30%、39%与 30%。根茎比在 R1 和 R3 期没有显著变化,但在 R5 和成熟期分别提升了 4%和 6%($P=0.06$)(表 1-1)。

表 1-1 大气 CO_2 浓度升高对地上部生物量、根系生物量、根冠比的影响

CO_2浓度 (μmol/mol)	年份	R1 地上部 (g/m²)	R1 根系 (g/m²)	R1 根/冠	R3 地上部 (g/m²)	R3 根系 (g/m²)	R3 根/冠	R5 地上部 (g/m²)	R5 根系 (g/m²)	R5 根/冠	收获 地上部 (g/m²)	收获 根系 (g/m²)	收获 根/冠
415	2009	94.4	13.2	0.139	455.2	71.8	0.158	564.73	75.4	0.134	764.5	76.4	0.099
415	2011	169.8	45.0	0.265	427.6	55.0	0.129	452.73	67.4	0.149	566.7	74.4	0.057
550	2009	119.4	15.2	0.127	466.4	76.6	0.166	732.8	105.2	0.144	884.0	91.7	0.104
550	2011	194.6	52.4	0.269	683	88.6	0.129	636.4	93.2	0.150	672.3	105.0	0.062
P值	CO_2	0.05	0.01	0.34	0.00	0.02	0.13	0.00	0.01	0.04	0.00	0.02	0.06
P值	年份	0.00	0.00	0.00	0.00	0.01	0.04	0.04	0.02	0.36	0.00	0.01	0.01
P值	交互	0.23	0.89	0.43	0.13	0.38	0.67	0.92	0.12	0.55	0.13	0.38	0.45

第二部分 大气 CO_2 浓度升高对主要 C_3 豆科作物的影响

1.3.2 氮、磷、钾浓度

大气 CO_2 浓度升高下植株地上部的氮浓度在 R1 阶段显著降低,但根系氮浓度与整株氮吸收没有显著变化。在 R3、R5 阶段与成熟期,地上部与根系氮浓度没有明显变化,但是整株氮吸收分别显著增加了 30.7%、34.2% 与 25.8%(表 1-2)。

表 1-2 大气 CO_2 浓度升高对大豆不同器官氮浓度的影响(%)

CO_2 浓度 ($\mu mol/mol$)	年份	生育时期							
		R1		R3		R5		成熟期	
		地上部 (%N)	根系 (%N)	地上部 (%N)	根系 (%N)	地上部 (%N)	根系 (%N)	地上部 (%N)	根系 (%N)
415	2009	6.10	2.24	2.73	0.94	2.39	0.67	1.98	0.50
	2011	2.56	1.09	2.76	0.89	2.96	0.96	2.01	0.51
550	2009	5.26	2.26	2.47	0.96	2.39	0.85	2.09	0.57
	2011	2.45	1.08	2.97	0.96	2.89	0.75	2.19	0.59
Anova P 值	CO_2	0.00	0.98	0.61	0.35	0.83	0.82	0.08	0.43
	年份	0.00	0.00	0.00	0.61	0.00	0.21	0.57	0.31
	交互	0.01	0.85	0.01	0.71	0.74	0.03	0.82	0.21

在 R1 与 R5 阶段,地上部磷浓度在 CO_2 浓度升高条件下分别显著增加了 11% 与 16%,氮在 R3 与成熟期没有变化。全生育期根系磷浓度在 CO_2 浓度升高下均没有显著变化,整株磷吸收在 R1、R3、R5 与成熟期分别显著增加了 37.1%、56.4%、51.9% 与 31.5%(图 1-1,表 1-3)。

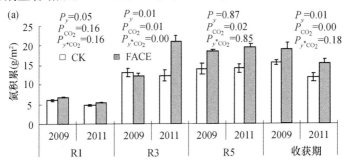

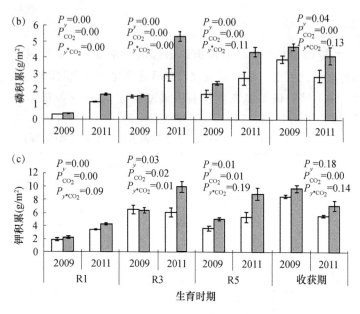

图 1-1 大气 CO_2 浓度升高对大豆氮、磷、钾吸收量的影响

(误差线表示标准差,P_y、P_{CO_2}、P_{y*CO_2} 分别表示 ANOVA 对年、CO_2 及年与 CO_2 的交互作用显著性的分析结果,下同。)

表 1-3 大气 CO_2 浓度升高对大豆不同器官磷浓度的影响

CO_2 浓度 (μmol/mol)	年份	生育时期							
		R1		R3		R5		成熟期	
		地上部 (%P)	根系 (%P)	地上部 (%P)	根系 (%P)	地上部 (%P)	根系 (%P)	地上部 (%P)	根系 (%P)
415	2009	0.32	0.17	0.29	0.20	0.27	0.18	0.48	0.27
	2011	0.55	0.47	0.62	0.47	0.50	0.55	0.43	0.49
550	2009	0.29	0.19	0.29	0.21	0.29	0.20	0.50	0.26
	2011	0.68	0.58	0.70	0.62	0.60	0.63	0.54	0.47

第二部分 大气 CO_2 浓度升高对主要 C_3 豆科作物的影响

续表

CO_2浓度 ($\mu mol/mol$)	年份	生育时期							
		R1		R3		R5		成熟期	
		地上部 (%N)	根系 (%N)	地上部 (%N)	根系 (%N)	地上部 (%N)	根系 (%N)	地上部 (%N)	根系 (%N)
P值	CO_2	0.04	0.10	0.16	0.07	0.05	0.21	0.19	0.86
	年份	0.00	0.00	0.00	0.00	0.00	0.00	0.73	0.02
	交互	0.01	0.21	0.19	0.12	0.16	0.46	0.35	0.24

大豆地上部与根系的钾浓度在大气 CO_2 浓度升高下没有显著变化，但钾吸收量显著增加（表1-4）。

表1-4 大气 CO_2 浓度升高对大豆不同器官钾浓度的影响

CO_2浓度 ($\mu mol/mol$)	年份	生育时期							
		R1		R3		R5		成熟期	
		地上部 (%K)	根系 (%K)	地上部 (%K)	根系 (%K)	地上部 (%K)	根系 (%K)	地上部 (%K)	根系 (%K)
415	2009	1.89	1.18	1.36	0.40	0.64	0.13	1.11	0.16
	2011	1.67	1.27	1.36	0.60	1.14	0.37	0.97	0.14
550	2009	1.75	1.08	1.30	0.42	0.66	0.16	1.09	0.14
	2011	1.85	1.35	1.36	0.69	1.32	0.42	1.04	0.16
P值	CO_2	0.95	0.92	0.75	0.41	0.13	0.06	0.75	0.52
	年份	0.14	0.03	0.16	0.01	0.00	0.00	0.06	0.65
	交互	0.01	0.24	0.46	0.67	0.21	0.58	0.54	0.69

1.3.3 酰脲浓度

大气 CO_2 浓度升高没有改变两年的 R1、R3 与 R5 期大豆上部完全展开叶的酰脲浓度（图1-2）。

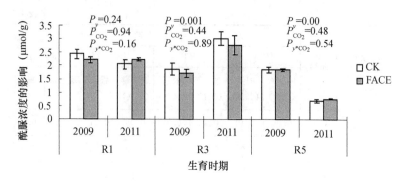

图 1-2 大气 CO_2 浓度升高对大豆叶片酰脲浓度的影响

1.3.4 氮、磷、钾在籽粒与地上部植株的积累

两年的平均数据表明,在大气 CO_2 浓度升高条件下,成熟期大豆籽粒氮、磷、钾的积累量分别显著增加了 23%、31%、26%,地上部氮、磷、钾的积累量分别显著增加了 25%、31%、19%(表 1-5)。

表 1-5 大气 CO_2 浓度升高大豆植株籽粒与地上部氮、磷、钾积累的影响

CO_2浓度 (μmol/mol)	年份	籽粒			地上部		
		氮 (g/m²)	磷 (g/m²)	钾 (g/m²)	氮 (g/m²)	磷 (g/m²)	钾 (g/m²)
415	2009	9.55	1.94	3.61	15.17	3.66	8.45
	2011	8.09	1.23	2.23	11.42	2.46	5.47
550	2009	11.55	2.55	4.58	18.50	4.44	9.63
	2011	10.13	1.59	2.75	14.77	3.63	6.98
Anova P值	CO_2	0.00	0.00	0.00	0.00	0.00	0.00
	Year	0.04	0.67	0.35	0.03	0.04	0.12
	Inter	0.43	0.79	0.91	0.21	0.16	0.17

1.4 讨论

大气 CO_2 浓度升高条件下,前人对水稻(Roy et al.,2012)、稗草

第二部分 大气 CO_2 浓度升高对主要 C_3 豆科作物的影响

(Zeng et al.,2011)、枫树以及白蜡树(Kanowski,2010)氮、磷、钾吸收利用的响应特征已有研究。本研究表明,大气 CO_2 浓度升高显著降低了开花初期大豆植株地上部氮浓度,但是对结荚期、灌浆期与成熟期没有显著影响。另外,Rogers 等(2010)的研究表明在大气 CO_2 浓度升高条件下大豆叶片氮含量在生育前期显著增加,但在中期与收获期没有变化。这应该归因于大气 CO_2 浓度升高引起植株的氮固定量与需求量的增加。然而,上层完全展开叶的酰脲(尿囊素与尿囊酸)浓度没有发生明显变化,这与美国 FACE 条件下的研究结果一致(Ainsworth et al.,2007)。前人研究表明,酰脲浓度与根瘤氮固定活力正相关,进而影响大豆对 N_2 的固定(Serraj et al.,2003;Serraj et al.,2010;Fang et al.,2008)。本研究中,酰脲浓度在大气 CO_2 浓度升高下没有明显变化,前人关于同一大豆品种中黄 35 的研究中植株固氮量也没有显著变化(Lam et al.,2012b),证明其氮固定能力没有提升。前人研究表明,土壤磷可利用性限制了三叶草、鹰嘴豆、紫花豌豆与苜蓿植株的氮固定(Edwards et al.,2010;Lam et al.,2012a)。然而,本研究中磷供应充足,可满足大气 CO_2 浓度升高下植株的生长与需求量,因而根系氮固定没有被磷的有效性限制。同时,氮固定量也没有显著增加,植株体内增加的氮积累量来自于土壤中的氮素。

前人用体内碳积累量增加对氮的稀释作用来解释大气 CO_2 浓度升高下植物组织内氮浓度降低的原因。然而,本研究中植株各器官氮浓度仅在生育前期降低,在中后期升高,表明"稀释作用"并不能完全解释本研究的试验结果。Bloom 等(2010)的研究表明,大气 CO_2 浓度升高限制了小麦与拟南芥中硝酸盐向有机氮的转化。但同样不能完全解释本研究中开花初期氮浓度降低的原因。所以我们认为,大豆开花初期氮需求与氮同化之间的不协调在较高的氮吸收量与发达的根系下被缓解。根茎比的增加也证明了这一点。

大豆植株磷浓度在 CO_2 浓度升高下增加(开花初期与灌浆期)或不变(结荚初期与收获期),但钾浓度没有变化。Loladze(2014)通过 meta 分析表明,大气 CO_2 浓度升高减少了多种矿质元素浓度,

但增加了 C_3 植物体内非结构性碳水化合物浓度。然而,一系列研究表明,大气 CO_2 浓度升高对豆科与非豆科植物体内的磷、钾的浓度没有或有积极影响(Kanowski,2010;Edwards et al.,2010;Li et al.,2015;Pérez-López et al.,2015),这与本研究的结果一致。所以,未来大气 CO_2 浓度升高背景下磷、钾元素吸收与代谢的生理机制亟待进一步研究。

大气 CO_2 浓度升高虽然对籽粒和地上部氮、磷、钾浓度的影响不明显,但由于增加了生物量积累,所以对氮、磷、钾的积累量也有促进作用。这与大气 CO_2 浓度升高下绿豆(Li et al.,2015)与水稻(Zeng et al.,2011)氮、磷、钾吸收的结果一致。大豆可能在大气 CO_2 浓度升高下吸收较多的营养物质。大气 CO_2 浓度升高对作物生长的影响受土壤营养条件与作物需求之间的动力学关系影响(Zeng et al.,2011)。Satapathy 等(2015)在 OTC 条件下的研究表明,土壤中氮的可利用性降低,因而印度亚热带水稻生产中需要施入更多的营养元素供给,以减少气候变化带来的危害。

为补偿籽粒携带走的磷、钾元素,不论秸秆是否被翻入地下,未来都需要在土壤中施入更多的磷、钾肥用以补偿。在大气 CO_2 浓度升高下,如果大豆固氮量增加,土壤中被施入较少的氮肥就可以满足植株生长需求。然而,中黄 35 的固氮量并没有增加(Lam et al.,2012b),所以仍需要将更多的氮肥施入土壤中以避免氮素亏缺。未来仍需要进一步进行不同土壤环境中不同类型大豆品种的适应性试验,以制定大气 CO_2 浓度升高环境下合理的氮、磷、钾管理措施。

1.5 结论

在不同生育时期,大气 CO_2 浓度升高对大豆植株氮、磷、钾浓度的影响存在差异,但在收获期没有明显差异。在收获期,大气 CO_2 浓度升高显著提高了地上部器官氮、磷、钾的积累量,所以未来需要施用更多的氮、磷、钾肥以维持大豆的生产活动。

第 2 章 大气 CO_2 浓度升高与干旱对大豆光合特性与产量的影响

2.1 引言

大气 CO_2 浓度升高、温度升高与降水格局发生改变被认为是对农业生态系统产生深远影响的主要气候因子(IPCC,2013;Leakey et al.,2009;Shao et al.,2015)。大气 CO_2 浓度升高常常可以增加多种作物的生产力与产量(Gao et al.,2015;Han et al.,2015),但是在干旱环境下,作物的调节适应过程会对产量形成产生较大限制(Bragazza,2008;Jentsch et al.,2011)。前人研究表明,大气 CO_2 浓度升高可以促进干旱条件下光合效率与水分利用效率提高,包括 C_3(Qiao et al.,2010)与 C_4 植物(Vu et al.,2009;Allen et al.,2011),这可以对干旱胁迫下作物生长与产量形成的抑制形成一定补偿(Zinta et al.,2014)。

大量研究表明,大气 CO_2 浓度升高对大豆叶片的光合能力、水分利用效率(气孔导度 G_s 降低)与产量均有促进作用(Hao et al.,2012;Morgan et al.,2005)。大气 CO_2 浓度升高条件下,WUE 的提高表明大豆的抗旱性可能会增强。然而,Li 等(2013a)的研究表明,大气 CO_2 浓度升高对正常水分条件下大豆 WUE 与生长的提高幅度大于水分胁迫条件。这表明,大气 CO_2 浓度升高不能较大程度上补偿干旱胁迫对大豆植株生长与籽粒产量形成造成的伤害。抗氧化能力、可溶性糖与叶绿素荧光特性被认为是干旱条件下反应植物抗旱性的重要指标(Razavi et al.,2008;Hatata et al.,2013)。前人研究表明,干旱胁迫会降低草莓植株的 PSⅡ光化学量子产量($\Phi_{PSⅡ}$)(Razavi et al.,2008)、小麦的 SOD(Hatata et al.,2013),但是增加了小麦的 POD(Hatata et al.,2013)与多种植物体内可溶性糖含量(Rosa et al.,2009)。

然而,鲜有对大气 CO_2 浓度升高环境中干旱胁迫对大豆抗氧化能力、可溶性糖与叶绿素荧光特性的适应特征进行研究。这对深入了解大气 CO_2 浓度升高环境下,大豆对干旱的适应机制有重要作用。本研究的目的是阐明大气 CO_2 浓度升高与干旱条件下,大豆植株的光合特征、抗氧化能力、叶绿素荧光特性、可溶性糖含量与 WUE 的变化规律。我们假设:(1)大气 CO_2 浓度升高可以缓解干旱胁迫对大豆植株产生的光合抑制与水分散失;(2)大气 CO_2 浓度升高可通过增强大豆植株的抗氧化能力与可溶性糖含量提高抗旱性。探明以上内容对解释大气 CO_2 浓度升高下大豆植株的抗旱适应性具有重要作用。

2.2 试验设计与方法

本试验于 2013—2014 年在山西农业大学人工模拟控制 CO_2 浓度的开顶式气室(OTC)进行。气室长 6 m,宽 4 m,高 3 m,顶开口呈八角形。供试品种为抗旱性较强的长旱 58。OTC 气室内地面中部安装带风扇的 CO_2 发散器,地面埋有与控制室 CO_2 钢瓶相连通的 CO_2 气体的管道,各气室安装 CO_2 浓度传感器及水分测定器,可随时监测室内 CO_2 浓度,高 CO_2 浓度室内 CO_2 气体由减压阀控制的压缩 CO_2 钢瓶供给。处理期间的 CO_2 浓度由钢瓶流出,进出温室中由通气小管散开,风扇运行,保证室内 CO_2 浓度均匀,设置高 CO_2 浓度处理的目标浓度,检测仪连续监测,当室内 CO_2 浓度低于目标浓度时,自动将 CO_2 注入室内,使其浓度保持在(560 ± 20 $\mu mol/mol$),通气时间 08:00—18:00,10 分钟记录一次数据,正常 CO_2 浓度室内只通入空气,不通入 CO_2。试验材料为大豆中黄 35。

大豆于 5 月 5 日播种于塑料种植箱中,箱长 60 cm,宽 40 cm,高 35 cm,在箱内部依次铺沙土、壤土共 28 cm 深。每个箱子播种 10 穴,每穴 2 粒种子,出苗后留 1 株健壮的幼苗。试验设置两个水分水平:(1)正常供水:土壤含水量为田间最大持水量的 80%;(2)干旱胁迫:土壤含水量为田间最大持水量的 45%。在每天 08:30 用便携式土壤水分测定仪(KZSF,中国)测定土壤含水量,将其控制在目标范围内。

于大豆生长的关键生育期测定气体交换参数(LI-6400型光合作用系统,LI-COR,美国)、叶绿素荧光特性(PAM-2500便携式荧光调制解调器,WALZ,德国)、POD(联苯胺比色法)、SOD(氮蓝四唑法)、MDA(硫代巴比妥酸法),于成熟期测定产量。

2.3 结果与分析

2.3.1 P_n 与气体交换参数

大气 CO_2 浓度升高下 P_n 在 2013 年和 2014 年平均升高了 44.1%,WUE 升高了 115.6%,T_r 降低了 16.5%,但 G_s 在两年均没有明显变化。在干旱条件下,大气 CO_2 浓度升高下 P_n 在开花期显著提高了 99.5%,灌浆期仅提高了 1.5%。在 2013 年和 2014 年,干旱使 P_n 降低了 44.3%,G_s 降低了 61.2%,C_i 降低了 8.3%,T_r 降低了 47.2%,但是 WUE 升高了 49.2%。CO_2 与 WUE 的交互作用对 WUE 有显著提升作用($P<0.01$),但对 P_n、G_s 与 T_r 没有显著影响。大气 CO_2 浓度升高加强了干旱对 WUE 的提升作用。交互作用 (5.79 = 9.35 − 3.56) 对 WUE 的提升作用显著高于单因素 (Drought:0.33 = 3.89−3.56;EC:1.75 = 5.31−3.56)(表 2-1)。

表 2-1 大气 CO_2 浓度升高与干旱对大豆第一片完全展开叶气体交换参数的影响

年份	生育期	水分条件	处理	净光合速率 (P_n)[μmol/(m²·s)]	气孔导度 (G_s)[mol(H_2O)/(m²·s)]	蒸腾速率 (T_r)[mmol(H_2O)/(m²·s)]	水分利用效率(WUE)[μmol(CO_2)/mmol(H_2O)]
2013	盛花期	正常	CK	13.84 c	0.24 bc	3.18 b	4.48 d
			EC	17.55 b	0.20 c	2.97 b	5.94 c
		干旱	CK	6.91 de	0.07 d	1.62 c	4.77 cd
			EC	10.63 c	0.08 d	1.47 c	7.22 b

续表

年份	生育期	水分条件	处理	净光合速率 (P_n)[μmol/($m^2 \cdot s$)]	气孔导度(G_s)[mol(H_2O)/($m^2 \cdot s$)]	蒸腾速率(T_r)[mmol(H_2O)/($m^2 \cdot s$)]	水分利用效率(WUE)[μmol(CO_2)/mmol(H_2O)]
2013	灌浆期	正常	CK	7.58 de	0.11 d	2.96 b	2.67 fg
		正常	EC	10.31 c	0.08 d	2.15 bc	4.81 d
		干旱	CK	11.00 c	0.20 c	4.77 a	2.33 g
		干旱	EC	10.18 cd	0.08 d	2.18 bc	4.51 d
2014	盛花期	正常	CK	17.46 b	0.32 ab	4.62 a	3.79 e
		正常	EC	24.19 a	0.41 a	5.00 a	4.85 d
		干旱	CK	7.55 de	0.08 d	1.57 c	4.92 cd
		干旱	EC	17.21 b	0.17 cd	2.83 b	6.29 c
	灌浆期	正常	CK	17.31 b	0.34 ab	5.29 a	3.32 ef
		正常	EC	25.34 a	0.31 ab	4.65 a	5.63 c
		干旱	CK	4.93 e	0.07 d	1.52 c	3.55 ef
		干旱	EC	5.97 de	0.01 e	0.32 d	19.36 a
P值	年份			0.00	0.00	0.00	0.00
	时期			0.00	0.00	0.66	0.07
	[CO_2]			0.00	0.42	0.00	0.00
	干旱			0.00	0.00	0.00	0.00
	年份 * 时期			0.48	0.30	0.00	0.00
	年份 * [CO_2]			0.00	0.02	0.01	0.00
	年份 * 干旱			0.00	0.00	0.00	0.00
	时期 * [CO_2]			0.00	0.00	0.00	0.00
	时期 * 干旱			0.59	0.02	0.13	0.00
	[CO_2] * 干旱			0.08	0.58	0.30	0.00
	年份 * 时期 * [CO_2]			0.67	0.17	0.74	0.00
	年份 * 时期 * 干旱			0.00	0.00	0.00	0.00
	时期 * [CO_2] * 干旱			0.00	0.15	0.02	0.00
	年份 * 时期 * [CO_2] * 干旱			0.33	0.73	0.30	0.00

2.3.2 叶绿素荧光特性

大气 CO_2 浓度升高条件下，PSⅡ最大光化学量子产量（F_v/F_m）、PSⅡ有效光化学量子产量（F_v'/F_m'）与非光化学淬灭系数（NPQ）没有显著变化，但PSⅡ实际光化学效率（$\Phi_{PSⅡ}$）显著提高了17.1%，光化学淬灭系数（qP）显著提高了27.0%。在干旱条件下，大气 CO_2 浓度升高下 $\Phi_{PSⅡ}$ 在2013年和2014年平均提升了16.0%，qP提升了16.8%。干旱下 F_v/F_m 降低了1.1%，F_v'/F_m' 降低了17.7%，$\Phi_{PSⅡ}$ 降低了20.5%，以及qP降低了10.1%，但NPQ升高了44.8%。CO_2 与干旱对 F_v/F_m、F_v'/F_m' 与NPQ没有显著影响，但对 $\Phi_{PSⅡ}$ （$P<0.05$）与qP（$P<0.01$）有显著影响。大气 CO_2 浓度升高减轻了干旱对 $\Phi_{PSⅡ}$ 与qP的抑制作用（表2-2）。

表2-2 大气 CO_2 浓度升高与干旱对大豆第一片完全展开叶叶绿素荧光特性的影响

年份	时期	水分条件	处理	F_v/F_m	F_v'/F_m'	$\Phi_{PSⅡ}$	qP	NPQ
2013	盛花期	正常	CK	0.80 abc	0.56 a	0.44 ab	0.78 bc	1.20 ef
			EC	0.82 a	0.59 a	0.49 a	0.82 bc	1.28 e
		干旱	CK	0.81 ab	0.43 b	0.39 b	0.75 c	1.96 c
			EC	0.81 ab	0.58 a	0.44 ab	0.93 ab	1.14 ef
	灌浆期	正常	CK	0.78 cd	0.48 b	0.32 c	0.65 cd	1.39 de
			EC	0.73 e	0.38 bc	0.39 b	1.14 a	1.28 def
		干旱	CK	0.78 cd	0.45 b	0.32 c	0.77 abc	1.60 d
			EC	0.78 cd	0.43 b	0.39 b	0.78 bc	1.49 de
2014	盛花期	正常	CK	0.80 bc	0.53 ab	0.30 c	0.57 de	1.61 cde
			EC	0.82 a	0.58 a	0.42 ab	0.72 c	1.37 de
		干旱	CK	0.76 de	0.41 b	0.23 d	0.55 ef	3.37 a
			EC	0.80 abc	0.32 c	0.25 cd	0.64 cd	2.10 bc
	灌浆期	正常	CK	0.81 ab	0.59 a	0.28 cd	0.47 f	1.05 fg
			EC	0.81 ab	0.46 b	0.24 d	0.51 ef	1.74 c

续表

年份	时期	水分条件	处理	F_v/F_m	F_v'/F_m'	Φ_{PSII}	qP	NPQ
2014	灌浆期	干旱	CK	0.78 cd	0.40 b	0.12 f	0.30 h	2.29 b
			EC	0.78 cd	0.42 b	0.15 e	0.34 g	1.86 c
P值			年份	0.10	0.10	0.00	0.00	0.00
			时期	0.00	0.00	0.00	0.00	0.04
			[CO_2]	0.42	0.46	0.05	0.05	0.38
			干旱	0.03	0.00	0.00	0.00	0.00
			年份 * 时期	0.00	0.00	0.80	0.00	0.01
			年份 * [CO_2]	0.00	0.07	0.28	0.87	0.00
			年份 * 干旱	0.00	0.00	0.00	0.23	0.00
			时期 * [CO_2]	0.00	0.00	0.03	0.33	0.34
			时期 * 干旱	0.09	0.01	0.83	0.00	0.04
			[CO_2] * 干旱	0.12	0.06	0.04	0.00	0.66
			年份 * 时期 * [CO_2]	0.68	0.05	0.26	0.00	0.02
			年份 * 时期 * 干旱	0.15	0.85	0.86	0.81	0.15
			stage * [CO_2] * 干旱	0.02	0.02	0.68	0.09	0.02
			年份 * 时期 * [CO_2] * 干旱	0.06	0.00	0.05	0.03	0.00

注：表中 F_v/F_m 为 PSⅡ最大光化学量子产量；F_v'/F_m' 为 PSⅡ有效光化学量子产量；Φ_{PSII} 为 PSⅡ实际光化学效率；qP 为光化学淬灭系数；NPQ 为非光化学淬灭系数。

2.3.3 过氧化物酶（POD）、超氧化物歧化酶（SOD）、丙二醛（MDA）与可溶性糖浓度

大气 CO_2 浓度升高对大豆叶片的 POD 与 MDA 没有显著影响，但 SOD 提升了 13.5%，可溶性糖提升了 26.9%。干旱对 SOD 与可

溶性糖没有显著影响,但是 POD 显著提升了 22.4%,MDA 显著提升了 23.8%。CO_2 与干旱的交互作用对 POD、SOD、MDA 和可溶性糖含量均没有显著影响(表 2-3)。

表 2-3 大气 CO_2 浓度升高与干旱对大豆过氧化物酶(POD)、超氧化物歧化酶(SOD)、丙二醛(MDA)和可溶性糖含量的影响

水分条件	处理	POD [μg/(gFW·min)]	SOD [μg/(gFW·h)]	MDA (mmol/gFW)	可溶性糖含量 (mmol/gFW)
正常	CK	22.53 a	146.15 b	0.22 b	5.57 c
	EC	19.90 b	163.14 a	0.20 c	7.27 ab
干旱	CK	27.40 a	145.09 b	0.27 a	6.49 b
	EC	24.56 a	167.31 a	0.25 b	8.04 a
P 值	[CO_2]	0.19	0.00	0.14	0.04
	干旱	0.03	0.78	0.01	0.25
	[CO_2]*干旱	0.95	0.63	0.88	0.91

2.3.4 植株形态指标

大气 CO_2 浓度升高下,大豆植株节数在 2013 年和 2014 年显著提升了 8.5%,茎直径提升了 22.6%,但是株高没有显著变化。干旱下,大豆株高降低了 32.4%,节数降低了 25.0%,茎直径降低了 35.7%。CO_2 与干旱的交互作用对株高、节数和茎直径没有显著影响(表 2-4)。

表 2-4 大气 CO_2 浓度升高与干旱对大豆形态指标和产量构成因素的影响

年份	水分条件	处理	株高 (cm)	节数	茎直径 (cm)	每株荚数	每荚粒数	百粒重 (g)
2013	正常	CK	65.47 b	10.34 c	0.47 b	22.83 b	1.92 ab	13.89 ab
		EC	61.41 b	9.74 cd	0.51 ab	25.38 ab	2.14 ab	11.89 b

续表

年份	水分条件	处理	株高(cm)	节数	茎直径(cm)	每株荚数	每荚粒数	百粒重(g)
2013	干旱	CK	47.36 c	7.86 e	0.35 d	16.26 cd	1.71 ab	13.42 ab
		EC	46.44 c	8.58 de	0.40 c	17.29 c	1.60 ac	10.91 bc
2014	正常	CK	67.21 b	12.43 b	0.42 c	23.49 ab	1.86 b	16.24 a
		EC	74.99 a	14.25 a	0.52 a	26.81 a	1.85 b	17.28 a
	干旱	CK	40.96 d	8.61 de	0.27 e	9.11 e	1.05 d	9.48 c
		EC	47.07 c	10.02 c	0.33 d	14.1 d	1.45 c	11.26 bc
P 值	年份		0.08	0.00	0.00	0.01	0.03	0.11
	[CO_2]		0.10	0.00	0.00	0.01	0.30	0.49
	干旱		0.00	0.00	0.00	0.00	0.00	0.00
	年份 * [CO_2]		0.00	0.00	0.10	0.23	0.56	0.01
	年份 * 干旱		0.00	0.00	0.03	0.00	0.33	0.00
	[CO_2] * 干旱		0.78	0.38	0.56	0.81	0.87	0.93
	年份 * [CO_2] * 干旱		0.35	0.10	0.37	0.38	0.14	0.61

2.3.5 生物量和产量

大气 CO_2 浓度升高下,每株植株的荚数在 2013 年和 2014 年平均显著提升了 16.6%,但是每荚粒数没有显著变化。干旱降低了每株荚数(42.4%)、每荚粒数(25.2%)和粒重(23.9%)。大气 CO_2 浓度升高下,地上部生物量显著升高了 17.5%,干旱条件下生物量的提升幅度(22.6%)高于正常水分条件(15.2%)。大气 CO_2 浓度升高下,大豆产量在两年平均显著升高了 17.7%,在干旱条件下显著提高了 29.6%,正常水分条件下显著提高了 12.9%。干旱下地上部生物量和产量分别平均降低了 44.9% 和 55.9%(表 2-5)。

表 2-5 大气 CO_2 浓度升高与干旱对大豆地上部生物量和产量的影响

年份	水分条件	处理	地上部生物量(g/m^2)	产量(g/m^2)
2013	正常	CK	491.67 c	200.00 b
		EC	542.67 c	207.00 b
	干旱	CK	322.00 d	125.33 c
		EC	335.33 d	127.33 c
2014	正常	CK	752.67 b	235.33 b
		EC	903.67 a	284.33 a
	干旱	CK	155.33 e	52.33 d
		EC	304.33 d	103.00 c
P 值	年份		0.00	0.57
	[CO_2]		0.01	0.05
	干旱		0.00	0.00
	年份 * [CO_2]		0.02	0.02
	年份 * 干旱		0.00	0.00
	[CO_2] * 干旱		0.53	0.50
	年份 * [CO_2] * 干旱		0.56	0.45

2.4 讨论

POD 和 SOD 是植物清除活性氧的第一道防线，MDA 是氧化胁迫的标志。干旱条件下，其活性与含量的改变被认为是水分胁迫下氧化还原态的重要标志(Moran et al.，1994；Schwanz et al.，2001)。可溶性糖作为营养和代谢信号分子，在非生物逆境下植物生理调节中起重要作用(Rosa et al.，2009)。本研究表明，大气 CO_2 浓度升高可以提升大豆叶片的 SOD 和可溶性糖含量，但降低了 POD 和 MDA。然而，CO_2 与干旱的交互作用对 POD、SOD、MDA 和可溶性糖含量均没有影响。这表明，大气 CO_2 浓度升高对干旱引起的植株防御系统的伤害没有缓解作用。

作为光合作用的底物，CO_2 和水对植物生长与产量形成都很重要。

CO_2 浓度升高具有通过增强光合作用并减弱蒸腾速率来提升作物瞬时 WUE 的潜力(Gao et al. ,2015;Sekhar et al. ,2020),这对帮助作物适应未来干旱环境有一定促进作用。本研究表明,大气 CO_2 浓度升高减缓了干旱对大豆叶片 WUE 造成的影响。干旱降低了 F_v/F_m、F_v'/F_m'、qP 和 Φ_{PSII},但升高了 NPQ,大气 CO_2 浓度升高减少了干旱对 Φ_{PSII} 和 qP 的抑制,表明大豆叶片的光合能力(Φ_{PSII} 和 qP)与 WUE 的提高,抗旱性增强。这与水稻(Baker et al. ,1997)、甘蔗(Joseph and Leon2009)和高粱(Vu et al. ,2009)的研究结果一致,但是与另一项关于大豆(Li et al. ,2013a)的研究不同。Li 等(2013a)的研究表明,大气 CO_2 浓度升高对正常水分条件下大豆 WUE 的提升幅度高于水分胁迫条件,且对干旱条件下大豆产量没有显著影响。另一方面,本研究表明,大气 CO_2 浓度升高减轻了干旱胁迫对地上作物生物量与产量的抑制作用,这与光合能力和 WUE 的提升有较大关系。但是本研究所采用的大豆品种属于高产品种,Li 等(2013a)的研究中所采用的大豆品种为高油品种,这可能是二者对 CO_2 浓度升高响应具有差异的原因。不同品种之间的生长、产量(Ziska et al. ,1999)、光合(Hao et al. ,2012)、氮固定(Lam et al. ,2012b)与氮、磷、钾吸收(Hao et al. ,2016)特性在前人的研究中也有报道。研究大豆生理和形态特征对大气 CO_2 浓度与干旱响应的种间差异对于探讨未来气候条件下的育种方向非常重要。

2.5 结论

干旱降低了大豆叶片 P_n、G_s、C_i、T_r、F_v/F_m、F_v'/F_m'、Φ_{PSII}、qP,但提升了 WUE、NPQ、POD、MDA,对 SOD 与可溶性糖含量没有显著影响。大气 CO_2 浓度升高促进了干旱对 WUE 的提升作用,缓解了干旱对 Φ_{PSII} 和 qP 的限制作用。所以,大气 CO_2 浓度升高通过增强光合性能和 WUE 提升了大豆的抗旱性。总之,大气 CO_2 浓度升高对中国北方干旱与半干旱地区大豆生产可能带来一定积极影响。本研究为未来大气 CO_2 浓度升高下干旱与半干旱地区大豆生产的可持续性及抗旱大豆品种的选育提供了一定理论依据。

第3章 大气 CO_2 浓度升高对绿豆元素吸收利用的影响

3.1 引言

大气 CO_2 浓度升高对多数植物的光合性能、器官发育，以及元素循环利用等多个过程造成深远影响(Pang et al., 2006; Gifford et al., 2000)。前人研究表明，在大气 CO_2 浓度升高背景下，水稻(Roy et al., 2012)、稗草(Zeng et al., 2011)、大豆(Ainsworth et al., 2007; Rogers et al., 2010)、铁线莲与巨盘木(Kanowski, 2010)、云杉(Li et al., 2013b)等植物对氮、磷、钾的吸收利用和器官分配差异较大，且在不同培养条件下也存在差异。在大田栽培条件下，大气 CO_2 浓度升高显著增强了水稻的吸氮量，但没有增加单位面积内氮积累量(Pang et al., 2006)与叶片氮浓度(Zeng et al., 2011)，稗草(Zeng et al., 2011)、铁线莲和巨盘木(Kanowski, 2010)的器官氮浓度均表现为降低。偏序钝叶草(Edwards et al., 2010)和水稻(Yang et al., 2007b)的研究表明，大气 CO_2 浓度持续升高环境下器官磷浓度显著升高。然而，水培条件下巨桉植株(Conroy et al., 2010)和盆栽条件下栓皮栎(Niinemets et al., 2010)植株的磷浓度明显降低。研究表明，钾浓度在巨盘木植株中降低，在麦珠子植株中没有明显变化。

大多数前人研究表明，植物对 CO_2 浓度持续升高响应的稳定性与土壤—植物系统中的氮素限制有关(Reich et al., 2006)。豆科植物可以通过氮固定增加氮素吸收，进而减少发生氮限制的可能性(Rogers et al., 2010)。FACE 条件下，大豆叶片氮含量在生育初期降低，但在中期对照没有明显变化(Rogers et al., 2010)。氨基酸与酰脲浓度在成熟叶片中没有明显变化，而在未完全成熟叶片中显著升高(Ainsworth et al., 2007)。但是，豆科植物对大气 CO_2 浓度升高的响应规律仍不十分明确，对氮、磷、钾的吸收与利用进行全面分析的研究较少。本研究旨在探明大气 CO_2 浓度升高条件下绿豆植株氮、磷、钾营养的利用特征,了解同化与代谢规律。

本研究利用 FACE 试验平台首次对大田条件下绿豆营养利用进行了研究。研究目的在于探明绿豆植株对大气 CO_2 浓度升高条件的响应特征,以及矿质营养吸收与利用效率。研究结果将有助于提供大气 CO_2 浓度持续升高背景下绿豆植株的氮、磷、钾利用策略,以及其对产量形成的影响,有利于对未来气候变化条件下豆科植株的适应机制进行深入了解。

3.2 试验设计与方法

本研究于2011年在位于北京市昌平区南部(40.13°N,116.14°E)的中国农业科学院昌平实验基地进行。田间开放式 CO_2 浓度控制通过中国农业科学院农业环境与可持续发展研究所研制的 Mini FACE 试验系统进行。Mini FACE 系统的设计同第2章。从绿豆出苗到收获全生育期期间,FACE 圈白天释放 CO_2,夜间不释放。CO_2 处理分别为 550 ± 60 $\mu mol/mol$(FACE)和 411 ± 20 $\mu mol/mol$(正常大气 CO_2 水平)2个水平。FACE 圈与对照之间其他田间管理措施一致。土壤类型属褐潮土,土壤全氮0.10%,有机质1.12%,有效磷 50 mg/kg,碱解氮 94.83 g/kg,pH 值 8.3。

于绿豆生长的关键生育期测定植株不同器官氮(凯氏定氮法)、磷(钼锑抗分光光度法)、钾(火焰光度法)含量。

3.3 结果分析

3.3.1 大气 CO_2 浓度升高对生物量的影响

在开花期初期,大气 CO_2 浓度升高对上部完全展开叶、其余叶片、茎、根与整株的生物量积累没有显著影响。在结荚期,绿豆植株的整株生物量在大气 CO_2 浓度升高的环境中增加了 14.2%($P=0.06$),但对全展开叶、其余叶片、茎与根的生物量积累没有显著影响(表3-1)。

第二部分 大气 CO_2 浓度升高对主要 C_3 豆科作物的影响

表3-1 大气 CO_2 浓度升高对绿豆不同器官生物量的影响

生育期	CO_2浓度 ($\mu mol/mol$)	第一片完全展开叶(g)	其余叶 (g)	茎 (g)	根 (g)	荚 (g)	籽粒 (g)	地上部 (g/m^2)
始花期	415	0.50±0.09	4.87±0.32	4.23±0.28	0.56±0.06			203.40±13.00
始花期	550	0.44±0.05	5.14±0.53	4.35±0.36	0.47±0.08			207.80±18.60
始花期	P值	0.57	0.68	0.81	0.38			0.85
结荚期	415	1.11±0.10	5.64±0.24	7.08±0.61	1.02±0.12	7.76±0.60		452.20±13.60
结荚期	550	1.33±0.10	6.87±0.55	7.83±0.18	1.06±0.08	8.73±0.68		516.40±20.4
结荚期	P值	0.21	0.11	0.31	0.79	0.34		0.06
成熟期	415		9.72±0.68	12.78±3.56	2.09±0.01	11.94±0.80	7.99±0.44	890.40±17.2
成熟期	550		10.29±0.83	14.51±0.95	2.25±0.13	14.99±0.61	10.05±0.22	1041.6±41.8
成熟期	P值		0.63	0.19	0.32	0.04	0.01	0.03

在成熟期,大气 CO_2 浓度升高条件下,荚、籽粒和整株生物量分别增加了 25.5%、25.8% 和 16.9%,但对叶、茎和根系没有显著影响。大气 CO_2 浓度升高对绿豆根系的根瘤数目没有显著影响,但使开花初期和结荚期每株根瘤菌的重量分别显著增加了 48.1% 和 30.1%(图 3-1、图 3-2)。

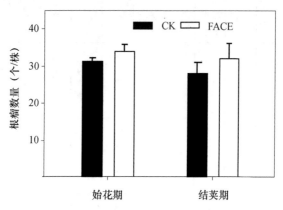

图 3-1 大气 CO_2 浓度升高对绿豆根瘤数量的影响

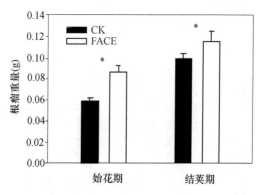

图 3-2 大气 CO_2 浓度升高对绿豆根瘤重量的影响(g)

第二部分 大气 CO_2 浓度升高对主要 C_3 豆科作物的影响

3.3.2 大气 CO_2 浓度升高对 N、P、K 含量的影响

在开花期初期,大气 CO_2 浓度升高对上部完全展开叶、其余叶片、茎、根和整株的氮浓度没有显著影响。在成熟期,绿豆植株的茎和整株氮浓度在大气 CO_2 浓度升高的环境中分别降低了 13.9% 和 4.4%,但是全展开叶、根、荚和籽粒的氮浓度没有显著变化(表 3-2)。

表 3-2 大气 CO_2 浓度升高对绿豆不同器官氮浓度和氮积累量的影响

生育期	CO_2 浓度 ($\mu mol/mol$)	第一片完全展开叶 (%N)	其余叶 (%N)	茎 (%N)	根 (%N)	荚 (%N)	籽粒 (%N)	地上部 (%N)	氮积累量 (g/m^2)
始花期	415	3.75±0.08	3.15±0.01	1.16±0.06	1.09±0.09			2.24±0.03	4.56±0.36
始花期	550	3.63±0.13	3.15±0.06	1.18±0.02	1.05±0.10			2.25±0.03	4.67±0.37
始花期	P值	0.47	0.97	0.70	0.77			0.78	0.84
结荚期	415	2.77±0.15	2.72±0.08	0.82±0.24	0.63±0.05	2.44±0.08		1.94±0.05	8.77±0.18
结荚期	550	2.74±0.12	2.66±0.05	0.85±0.26	0.75±0.08	2.38±0.13		1.94±0.05	9.98±0.29
结荚期	P值	0.89	0.56	0.47	0.24	0.68		0.92	0.02
成熟期	415		2.21±0.09	0.79±0.01	0.98±0.06	0.54±0.04	3.88±0.07	1.60±0.01	14.23±0.23
成熟期	550		2.05±0.07	0.68±0.04	0.92±0.03	0.54±0.23	3.84±0.70	1.53±0.02	15.91±0.64
成熟期	P值		0.24	0.04	0.43	0.89	0.76	0.03	0.07

在结荚期和成熟期,每平方米绿豆氮素吸收量在大气 CO_2 浓度升高的环境中分别增加了 13.9% 和 11.6%($P=0.07$),但开花初期没有显著变化。在开花初期,大气 CO_2 浓度升高下根系磷浓度显著增加了 13.2%,但叶片、茎和整株没有显著变化。在结荚期,大气 CO_2 浓度升高下绿豆下部叶片的磷浓度显著增加了 15.1%,但植株上部完全展开叶、茎、根、荚和整株没有显著变化(表 3-3)。

表 3-3 大气 CO_2 浓度升高对绿豆不同器官磷浓度和磷积累量的影响

生育期	[CO_2](μmol/mol)	第一片完全展开叶(%P)	其余叶(%P)	茎(%P)	根(%P)	荚(%P)	籽粒(%P)	地上部(%P)	磷积累量(g/m^2)
始花期	415	0.73±0.02	0.79±0.08	0.55±0.01	0.53±0.02			0.67±0.04	1.37±0.17
始花期	550	0.76±0.06	0.66±0.04	0.63±0.06	0.60±0.02			0.65±0.02	1.35±0.13
始花期	P 值	0.70	0.23	0.27	0.04			0.65	0.92
结荚期	415	0.52±0.05	0.53±0.01	0.70±0.08	0.64±0.15	0.66±0.04		0.63±0.01	2.85±0.12
结荚期	550	0.57±0.60	0.61±0.01	0.60±0.03	0.58±0.11	0.67±0.04		0.62±0.02	3.21±0.06
结荚期	P 值	0.58	0.00	0.29	0.79	0.81		0.76	0.06
成熟期	415		0.67±0.05	0.58±0.08	0.76±0.05	0.29±0.03	0.71±0.02	0.55±0.03	4.92±0.18
成熟期	550		0.64±0.06	0.57±0.05	0.53±0.40	0.35±0.02	0.77±0.03	0.56±0.01	5.80±0.27
成熟期	P 值		0.82	0.89	0.03	0.19	0.13	0.92	0.05

第二部分 大气 CO_2 浓度升高对主要 C_3 豆科作物的影响

在成熟期,根系磷浓度显著降低了 30.3%。大气 CO_2 浓度升高下叶、茎、荚、籽粒和整株的磷浓度在成熟期没有变化。每平方米植株的磷吸收量在成熟期显著增加了 21.7%,在开花初期和结荚期没有显著变化。

表 3-4 大气 CO_2 浓度升高对绿豆不同器官钾浓度与钾积累量的影响

生育期	CO_2浓度 (μmol/mol)	第一片完全展开叶 (%K)	其余叶 (%K)	茎 (%K)	根 (%K)	荚 (%K)	籽粒 (%K)	地上部 (%K)	钾积累量 (g/m^2)
始花期	415	1.47±0.07	1.76±0.07	3.62±0.00	1.80±0.07			2.52±0.02	5.12±0.36
始花期	550	1.70±0.11	1.78±0.15	3.70±0.54	2.07±0.17			2.59±0.09	5.36±0.32
始花期	P 值	0.17	0.91	0.21	0.21			0.48	0.65
结荚期	415	1.13±0.09	1.37±0.11	2.65±0.19	0.93±0.12	1.41±0.07		1.76±0.14	7.97±0.79
结荚期	550	1.11±0.07	1.51±0.07	2.87±0.17	1.15±0.10	1.35±0.05		1.83±0.06	9.47±0.57
结荚期	P 值	0.87	0.33	0.43	0.24	0.54		0.66	0.19
成熟期	415		1.13±0.06	1.70±0.11	0.93±0.12	1.37±0.11	1.17±0.04	1.35±0.01	12.03±0.20
成熟期	550		1.09±0.10	1.88±0.04	0.93±0.40	1.41±0.05	1.11±0.04	1.40±0.01	14.53±0.49
成熟期	P 值		0.75	0.19	0.99	0.75	0.35	0.06	0.01

大气 CO_2 浓度升高对上部叶片、其余叶片、茎、根、荚、籽粒和整株的钾浓度没有显著影响。大气 CO_2 浓度升高下,每平方米绿豆钾的吸收量在成熟期显著增加了 22.4%,开花初期和结荚期没有显著变化(表 3-4)。

3.3.3 氮、磷、钾吸收与利用效率

在成熟期,大气 CO_2 浓度升高下氮、磷、钾的籽粒吸收量分别显著增加了 25.85%、36.8% 和 20.4%,地上部吸收量分别显著增加了 11.6%、21.7% 和 22.4%。大气 CO_2 浓度升高对成熟期籽粒的氮、磷、钾利用率没有显著影响(图 3-3、表 3-5)。

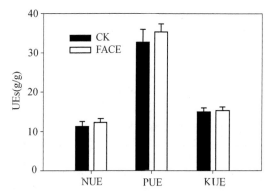

图 3-3 大气 CO_2 浓度升高对绿豆籽粒氮利用效率(NUE)、磷利用效率(PUE)、钾利用效率(KUE)的影响

表 3-5 大气 CO_2 浓度升高对绿豆籽粒与地上部氮磷钾积累的影响 (g)

CO_2 浓度 ($\mu mol/mol$)	籽粒			地上部		
	氮 (g)	磷 (g)	钾 (g)	氮 (g)	磷 (g)	钾 (g)
415	0.31±0.02	0.06±0.00	0.09±0.00	0.69±0.01	0.23±0.01	0.58±0.01
550	0.39±0.01	0.08±0.00	0.11±0.01	0.77±0.03	0.28±0.01	0.71±0.02
P 值	0.03	0.01	0.05	0.06	0.03	0.01

第二部分 大气 CO_2 浓度升高对主要 C_3 豆科作物的影响

3.4 讨论

大气 CO_2 浓度升高对开花初期和结荚期绿豆整株氮浓度没有显著影响,但降低了成熟期整株氮浓度。目前,大气 CO_2 浓度升高下植物组织中氮浓度降低的原因有 5 种解释,即"稀释""氮分配改变""需求量降低""蒸腾降低引发的氮吸收量降低"(Pang et al.,2006)和"硝酸盐同化被抑制"(Bloom et al.,2010)。本研究中,(1)植物体中碳水化合物增加引起氮的稀释并不能解释磷与钾含量无变化。(2)体内器官氮分配:前人研究表明,无论在氮素匮乏(PiaWalch-Liu et al.,2001)还是供氮充足(Gifford et al.,2000;Zong et al.,2013)的条件下,大气 CO_2 浓度升高下植株地上部氮分配显著降低,而根系氮分配显著增加。我们的研究表明,根系氮分配在大气 CO_2 浓度升高环境中并没有改变(未发表),Pang 等(2006)对水稻的研究也得出相似结论。(3)大气 CO_2 浓度升高增加了氮利用效率(NUE),降低了地上部氮的需求量(Li et al.,2003;Zong et al.,2014),导致 C_3 植物叶片氮浓度降低(Rogers et al.,2010)。本研究中,绿豆 NUE 在 CO_2 浓度升高环境中并没有升高。(4)气孔部分关闭与叶片蒸腾速率的降低会引起土壤中氮素集流与根系吸收量降低(Conroy et al.,2010;Drake,1997)。本研究表明,大气 CO_2 浓度升高下气孔开度降低,但是并没有数据证明其直接引发了氮素吸收速率的降低(Gao et al.,2015)。蒸腾速率的降低不是大气 CO_2 浓度升高下氮素吸收量减少的主要原因(Pang et al.,2006)。因而,"稀释""氮分配改变""需求量降低""蒸腾降低引发的氮吸收量降低"都不能完全解释成熟期绿豆氮浓度降低的原因。(5)大气 CO_2 浓度升高会限制硝酸盐同化为有机氮化合物,这可能是由于 CO_2 持续升高下 C_3 植物光合碳同化能力与生长速率降低所引起的(Bloom et al.,2010)。本研究表明,CO_2 浓度升高下绿豆的光合能力降低,特别是在结荚期(Gao et al.,2015)。所以硝酸盐同化受抑制可能是成熟期植物体内氮浓度降低的原因。

大气 CO_2 浓度增加降低了大豆生育前期的叶氮含量,但是对生育中期与收获期没有显著影响(Rogers et al.,2010)。这可能与植物在高 CO_2 环境中增加了对氮素的需求有关。本研究表明,大气 CO_2 浓度升高条件显著增加了开花初期和结荚期绿豆根瘤重量,与 Rogers 等(2010)对于大豆的研究结果一致。作为固氮植物,绿豆可以固定并吸收空气中的氮素,所以具有缓解由硝酸盐同化受抑制而引起的氮含量降低的潜力。前人研究表明,根瘤菌的固氮酶活性在绿豆开花之后增加,但在灌浆期降低(Mastrodomenico et al.,2013),这可能是氮浓度在开花初期与结荚期无变化,但在成熟期降低的原因。生物量在成熟期降低了 16.9%,降低幅度高于其余时期。在土壤磷含量较高条件下,大气 CO_2 浓度升高可以增加三叶草的固氮量,但对低磷土壤中氮素的固定量没有显著影响(Lam et al.,2012a;Edwards et al.,2010)。本研究中,播前土壤有效磷含量较高(50 mg/kg),对生育前期根瘤菌的形成与固氮具有一定的促进作用。随着植株生育期的推进与对土壤中磷素吸收量的增加,有效磷含量在绿豆生育后期降低。这可能是后期根系中磷含量较低的原因之一。成熟期氮固定量的降低可能与磷的有效性有关。大气 CO_2 浓度升高条件下,较高的生物量积累需要较高的氮素维持植株生长,但是由于硝酸盐同化受抑制,绿豆根系对氮的吸收量有限(Bloom et al.,2010),且根瘤菌没有从空气中固定更多氮素。这也许是成熟期植株氮含量较低的另外一个原因。

前人研究表明,在大气 CO_2 浓度升高环境中,水稻(Zeng et al.,2011;Yang et al.,2007b)、钝叶草(Edwards et al.,2010)和稗草(Zeng et al.,2011)体内磷浓度增加,但桤属(Niinemets et al.,2010)和巨盘木属(Kanowski,2010)没有显著变化。生长在大气 CO_2 浓度升高环境中的水稻与稗草(Zeng et al.,2011)钾浓度增加,但是巨盘木属(Kanowski,2010)显著降低,本研究表明,绿豆整株磷、钾浓度在全生育期均无显著变化,但根系与下部叶片磷浓度显著增加(开花初期),根系磷浓度在成熟期显著下降。这个差异可能与植株生长环境以及品种有关。大气 CO_2 浓度升高对磷与钾吸收代谢的影响机制需要进行进一步研究。

然而，大气 CO_2 浓度升高条件下，成熟期绿豆氮浓度降低，磷、钾浓度没有显著变化。植株氮、磷、钾的积累量与生物量显著增加。氮、磷、钾的籽粒含量与地上部含量在大气 CO_2 浓度升高条件下均显著增加。所以本研究表明，植物生物量积累与土壤营养消耗之间的营养动力学关系可能在未来发生变化，大气 CO_2 浓度升高条件下绿豆植株需要从土壤中吸收更多的营养元素。不论秸秆是否翻入土壤，向土壤中施用更多的磷、钾肥都是必要的。因为绿豆可以固定并吸收空气中的氮素，如果大气 CO_2 浓度升高条件下绿豆的固氮量增加，植株固定在土壤中的氮素将有可能抵消由于籽粒带走的氮素，因而将不需要向土壤中施用更多的氮素。然而，目前仍不十分清楚，需要通过获得不同的实验条件与不同品种的更多数据来制定更加有效的氮、磷、钾管理策略。

总之，大气 CO_2 浓度升高对绿豆氮利用效率（NUE）、磷利用效率（PUE）和钾利用效率（KUE）均没有显著影响。这与水稻的结论不同，有前人研究证明水稻的 NUE 增加（Roy et al.，2012），而 PUE（Yang et al.，2007b）降低。不同的生长条件和品种会影响大气 CO_2 浓度升高下植株对氮、磷、钾的利用，所以仍需要更多的试验证明。

3.5 结论

综上所述，大气 CO_2 浓度升高增加了成熟期绿豆整株生物量的积累，但降低了氮浓度，未改变磷、钾浓度。大气 CO_2 浓度升高也增加了开花期与结荚期绿叶根系根瘤的重量。氮固定能力在大气 CO_2 浓度升高条件下可能会增加，但是 NUE、PUE 和 KUE 没有发生改变。在成熟期，每株与每平方米氮、磷、钾的吸收量均显著增加，籽粒与地上部的氮、磷、钾积累量也显著增加。本研究表明，不同营养元素在植株体内的积累与土壤消耗中的动力学关系在未来 CO_2 浓度升高条件下将会改变，需要进行更多研究去制定未来气候变化背景下的施肥策略。

第4章 大气 CO_2 浓度升高对绿豆光合特性与产量的影响

4.1 引言

大气 CO_2 浓度升高会促进植物光合能力提升,进而增强光合产物供应、生物量积累与产量形成(Drake,1997;Long et al.,2004;Pérez-López et al.,2015)。总体上来讲,大气 CO_2 浓度升高增加了 Rubisco 的羧化速率,进而限制了 RuBP 氧化(Bowes,2003)。在大气 CO_2 浓度升高环境中,作物的响应因种类而异,主要原因与光合适应策略有关(Andrews et al.,2010)。光合生物化学与动力特征与库强和 Rubisco 的分子生物特性是降低大气 CO_2 浓度升高下叶片光合能力下降的原因(Bowes,2003;Hao et al.,2012)。

绿豆(*Vigna radiata* L.)是中国重要的传统作物,在全球范围内广泛种植。绿豆不仅具有较高的营养价值,并且有助于固定土壤空气中的氮素。绿豆籽粒是中国、印度、韩国、巴基斯坦、日本、泰国等东亚与东南亚国家的重要农产品,其籽粒蛋白质浓度高而脂肪浓度较低(Zhang et al.,2013)。前人研究表明,大气 CO_2 浓度升高背景下不同品种的大豆叶片均有出现光合适应的可能,但是在库强度较大的大豆品种中却没有出现(Hao et al.,2012)。桃树植株的氮固定对于维持糖分水平起到促进作用,进而有助于避免产生光合适应(Aranjuelo et al.,2014)。然而,目前无论在封闭式还是开放式的试验平台,关于绿豆植株对大气 CO_2 浓度升高的适应性研究都较少(Xue-Xia et al.,2007;Hao et al.,2014;Hao et al.,2012)。

前人研究表明,在大气 CO_2 浓度升高条件下,丛枝菌根的繁殖与菌丝长度显著增加(Yang et al.,2007a)。在封闭性的 CO_2 浓度升高模拟平台下,Yuan 等(2007)的研究表明,绿豆的根/茎在注射丛枝菌根菌后显著增加,大气 CO_2 浓度升高增加了绿豆的 P_n 与 WUE,同时降低了 G_s 与 T_r。也有研究表明,大气 CO_2 浓度升高降低了绿豆结荚期的 F_v/F_m(Hao et al.,2011)。大多数控制条件如盆栽研究的试验结果与开放条件下大田研究的结果有差异,可能与限制了大气 CO_2 浓度升高环境中根系的进一步扩大有关(Mcleod et al.,1999;Long et al.,2004)。

总体上来讲,大部分叶绿素 b 与 a 吸收光能,并将其用于光合碳同化。在这个过程中,类胡萝卜素保护反应中心免受多余的光能伤害,并有助于截获光能。所以,光合色素的变化与植株生理状况、生产力关系密切(Bernstein,1974;Blackburn,1998)。

大气 CO_2 浓度升高下不同作物的生理生态响应特征各异(Ainsworth et al.,2007;Hao et al.,2012)。本研究是第一个在开放式大气 CO_2 浓度模拟试验平台下,研究绿豆叶片的光合生理、叶绿素荧光特性与产量构成。本研究的目的在于揭示:(1)叶片的光合生理与叶绿素荧光特性在大气 CO_2 浓度持续升高条件下是否发生变化,以及其与光合参数之间的关系。(2)大气 CO_2 浓度升高是否增强了绿豆叶片的光合性能、生物量积累和产量形成。

4.2 试验设计与方法

本研究于 2011 年在位于北京市昌平区南部(40.13°N,116.14°E)的中国农业科学院昌平实验基地进行。田间开放式 CO_2 浓度控制通过中国农业科学院农业环境与可持续发展研究所研制的 Mini FACE 试验系统进行。Mini FACE 系统的设计同第 2 章。从大豆出苗到收获全生育期期间,FACE 圈白天释放 CO_2,夜间不释放。CO_2 处理分别为 550 ± 60 $\mu mol/mol$(FACE)和 411 ± 20 $\mu mol/mol$(正常大气 CO_2 水平)2 个水平。FACE 圈与对照之间其他田间管理措施一致。土壤类型属褐潮土,土壤全氮 0.10%,有机质 1.12%,有效磷 50 mg/kg,碱解氮 94.83 g/kg,pH 值 8.3。

于大豆生长的关键生育期测定气体交换参数(LI-6400型光合作用系统,LI-COR,美国)、叶绿素荧光特性(PAM-2500便携式荧光调制解调器,WALZ,德国)、叶绿素含量(乙醇丙酮静提法),于成熟期测定产量。

4.3 结果分析

4.3.1 光合色素浓度

在开花期,大气CO_2浓度升高较正常CO_2环境下叶片叶绿素a与总叶绿素浓度分别增加了10.9%和10.6%。但是叶绿素b浓度没有显著变化。在成熟期(R4),叶绿素a、b、类胡萝卜素和叶绿素a/b在CO_2浓度升高环境下均无显著变化(图4-1)。

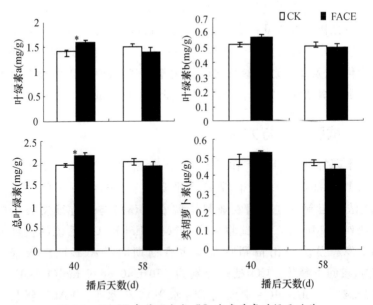

图4-1 FACE条件下大气CO_2浓度升高对绿豆叶片光合色素含量的影响

第二部分 大气 CO_2 浓度升高对主要 C_3 豆科作物的影响

4.3.2 P_n 与气体交换参数

CO_2 浓度升高环境下绿豆上部完全展开叶的 P_n 在开花期和成熟期分别较对照增加了 18.7% 和 7.4%。同时,G_s 较对照分别降低 19.2% 和 13.7%。T_r 无显著变化。因而,开花期和成熟期的 WUE 在大气 CO_2 浓度升高条件下分别提升了 38.9% 和 19.9%。进一步对光合特征曲线的分析表明,最大电子传递速率(J_{max})和最大羧化速率($V_{c,max}$)并未在大气 CO_2 浓度升高环境下升高(表 4-1)。

表 4-1 大气 CO_2 浓度升高对绿豆第一片完全展开叶气体交换特征的影响

播后天数(d)	处理	净光合速率 (P_n)[μmol(CO_2)/($m^2 \cdot s$)]	气孔导度 (G_s)[mol(H_2O)/($m^2 \cdot s$)]	蒸腾速率 (T_r)[mmol/($m^2 \cdot s$)]	水分利用效率(WUE)[μmol/mmol]	$V_{c,max}$ [μmol/($m^2 \cdot s$)]	J_{max} [μmol/($m^2 s \cdot$)]
40	CK	21.67±0.79	1.33±0.10	11.15±1.02	2.02±0.25	84.61±3.31	107.79±3.59
40	FACE	25.72±0.02	1.08±0.03	9.41±1.10	2.80±0.30	90.79±2.93	117.37±1.62
58	CK	30.63±0.52	1.22±0.08	9.29±0.46	3.32±0.21	112.40±4.23	142.27±14.12
58	FACE	32.90±0.65	1.05±0.04	8.30±0.36	3.98±0.16	110.90±4.02	142.12±8.84
P 值	生育时期	0.00	0.35	0.10	0.00	0.00	0.05
P 值	[CO_2]	0.00	0.02	0.13	0.01	0.26	0.24
P 值	生育时期*[CO_2]	0.16	0.54	0.66	0.75	0.17	0.23

4.3.3 叶绿素荧光特性

大气 CO_2 浓度升高下，F_v/F_m 较 CK 没有发生明显变化。但是，CO_2 浓度升高下 F_v'/F_m' 在播种 40 天和 58 天分别显著降低了 2.8 和 13.8%，同时，Φ_{PSII} 分别降低了 2.3% 和 34.5%。PSⅡ反应中心的开放程度在 CO_2 浓度升高下分别降低了 0.4% 和 23.2%。另一方面，非光化学淬灭系数（NPQ）分别增加了 12.8% 和 28.2%。大气 CO_2 浓度升高下，成熟期绿豆叶片叶绿素荧光特征参数的降低幅度大于开花期（表 4-2）。

表 4-2　大气 CO_2 浓度升高对绿豆第一片完全展开叶叶绿素荧光特征的影响

播后天数 (d)	处理	F_v/F_m	F_v'/F_m'	Φ_{PSII}	qP	NPQ
40	CK	0.83±0.00	0.58±0.01	0.38±0.01	0.65±0.02	1.26±0.07
40	FACE	0.83±0.01	0.56±0.02	0.37±0.02	0.65±0.01	1.42±0.10
58	CK	0.80±0.01	0.55±0.03	0.36±0.04	0.64±0.03	1.11±0.18
58	FACE	0.78±0.04	0.47±0.02	0.23±0.02	0.49±0.02	1.42±0.07
P 值	生育时期	0.07	0.00	0.00	0.00	0.97
P 值	[CO_2]	0.46	0.05	0.01	0.00	0.03
P 值	生育时期 * [CO_2]	0.57	0.20	0.02	0.00	0.21

注：表中 F_v/F_m 为 PSⅡ最大光化学量子产量；F_v'/F_m' 为 PSⅡ有效光化学量子产量；Φ_{PSII} 为 PSⅡ实际光化学效率；qP 为光化学淬灭系数；NPQ 为非光化学淬灭系数。

4.3.4 生物量、产量以及产量构成

总生物量和籽粒亩产量在 CO_2 浓度升高下分别升高了 11.6%

和 14.2%。粒数提升了 11.8%,与籽粒产量的提高成正相关关系。但是,大气 CO_2 浓度的升高对百粒重没有显著影响(图 4-2、表 4-3)。

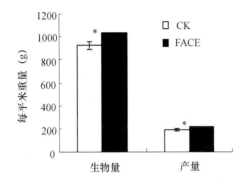

图 4-2 大气 CO_2 浓度升高对绿豆单位面积地上部生物量和产量的影响

表 4-3 大气 CO_2 浓度升高对绿豆产量构成因素的影响

处理	每株荚数(个)	每株粒数(个)	百粒重(g)
CK	15.93±1.31	7.63±0.16	6.57±0.26
FACE	18.13±0.48	8.53±0.15	6.50±0.19
P 值	0.19	0.02	0.86

4.4 讨论

前人研究表明,光合色素浓度的变化与光合生理特性、植物生产力显著相关。在最近的研究中,Jiang 等(2006)和 Zhao 等(2003)对 OTC 条件下大豆叶片光合色素的变化进行了分析,表明大气 CO_2 浓度的升高可以提高叶绿素 a、b、总叶绿素以及类胡萝卜素含量,但在 Hao 等(2012)的研究中未发生明显变化。黑云杉与红云杉的总叶绿素 a+b 含量在大气 CO_2 浓度升高环境下显著降低(John et al.,2007)。但是,本研究表明,大气 CO_2 浓度升高提高了绿豆开花期叶绿素浓度,但对成熟期没有显著影响。这些结论表明,光合色素浓度在大气 CO_2 浓度升高下的变化因植物种类和时期而异,这

与 C_3 植物叶绿素含量变化规律的差异较大(Seneweera et al.,2005)。在本研究中,大气 CO_2 浓度升高下绿豆叶片 P_n 在开花期处于最高水平。这个结果与叶绿素浓度以及气体交换参数相印证。大气 CO_2 浓度升高下,叶绿素浓度在开花期升高了 10.6%,但在结荚期没有明显变化。

前人研究表明,大气 CO_2 浓度升高下 C_3 植物的光合能力常常被提高,其中最大光合速率(A_{sat})在 FACE 条件下提高了 31%(Ainsworth et al.,2007)。前人研究表明,大气 CO_2 浓度升高环境下叶片光合功能的适应主要被 Rubisco 的含量与活性限制(Ainsworth et al.,2010)。本研究结果表明,虽然叶片 P_n 在 CO_2 浓度升高环境下显著增强,但 J_{max} 和 $V_{c,max}$ 并没有明显变化。

Φ_{PSII} 可以表征 PSII 的光化学效率(Tausz-Posch et al.,2013)。qP 反映了光响应中的非光化学响应过程,也可能导致非光化学耗散(NRD)(Rohacek,2002)。F_v'/F_m' 作为光适应下的 PSII 潜在效率,NPQ 表明叶片热耗散能力(Myers et al.,2010)。前人研究表明,当库器官受抑制时,NPQ 显著增加(Myers et al.,2010)。本研究中,当植物在高 CO_2 环境中,F_v'/F_m'、Φ_{PSII} 和 qP 均显著下降,且成熟期下降幅度高于开花期。这些结论进一步证实,在大气 CO_2 浓度升高条件下,成熟期绿豆的光合能力降低,且库强降低。同时,NPQ 显著增加,更多的能量被耗散,而不是被光化学过程利用。然而,在叶片衰老过程中,热耗散的增加以及其引发的碳同化限制对衰老的贡献率仍不十分清楚。大气 CO_2 浓度升高条件下,P_n 通常会被增强。同时,在成熟期,叶片 P_n 低于开花期,这与叶绿素荧光参数与气体交换参数的结果相一致。前人研究表明,源库关系影响了大气 CO_2 浓度升高下叶片的光合适应特征(Ainsworth et al.,2010;Hao et al.,2012)。本研究结果表明,绿豆植株的库器官在 CO_2 浓度升高下有一定的提升空间。CO_2 浓度升高下绿豆生物量和产量分别提高了 11.6% 和 14.2%。产量提升的主要原因为粒数的增加,单粒重和荚数并未显著变化。

前人研究表明,G_s 在 CO_2 浓度升高下显著降低,但是并未发生长期的气孔限制（Leakey et al.,2006）。G_s的降低可以提高WUE。本研究中，CO_2浓度升高下G_s降低，但T_r并未发生明显变化，所以绿豆水分利用效率的大幅度提高主要是由P_n升高引起的，表明CO_2浓度升高会增强绿豆的抗旱能力。

4.5 结论

CO_2浓度升高条件下，开花期绿豆的叶绿素浓度、上层完全展开叶的P_n、WUE和NPQ显著增加，但是F_v'/F_m'、$\Phi_{PSⅡ}$和qP降低。尽管降低的叶绿素荧光参数引起了光合限制，P_n、生物量和产量仍显著提高。产量的提高主要是由粒数增加引起的，而不是粒重和荚数。本研究表明，如果库强可以维持或增强，绿豆产量在未来CO_2浓度提升环境下会得到提升。

参考文献

郝兴宇,韩雪,李萍,等,2011.大气CO_2浓度升高对绿豆叶片光合作用及叶绿素荧光参数的影响[J].应用生态学报,22(10):2776-2780.

赵天宏,史奕,黄国宏,2003.CO_2和O_3浓度倍增及其交互作用对大豆叶绿体超微结构的影响[J].中国应用生态学报,14(12):2229-2232.

Ainsworth E A, Alistair R, Leakey A D B, et al, 2007. Does elevated atmospheric [CO_2] alter diurnal C uptake and the balance of C and N metabolites in growing and fully expanded soybean leaves? [J]. Journal of Experimental Botany (3):579-591.

Ainsworth E A, Rogers A, 2010. The response of photosynthesis and stomatal conductance to rising [CO_2]: mechanisms and environmental interactions [J]. Plant, Cell and Environment, 30 (3):258-270.

Allen L H, Kakani V G, Vu J C V, et al, 2011. Elevated CO_2 increases water use efficiency by sustaining photosynthesis of water-limited maize and sorghum [J]. Journal of Plant Physiology, 168:1909-1918.

Andrews M, Raven J A, Sprent J I, 2010. Environmental effects on dry matter partitioning between shoot and root of crop plants: relations with growth and shoot protein concentration [J]. Annals of Applied Biology,138 (1):57-68.

Aranjuelo I, Cabrerizo P M, Aparicio-Tejo P M, et al, 2014. Unravelling the mechanisms that improve photosynthetic performance of N_2-fixing pea plants exposed to elevated [CO_2] [J]. Environmental and Experimental Botany,99: 167-174.

Baker J, Hartwell A L, Boote K, et al, 1997. Rice responses to drought under carbon dioxide enrichment. 1. Growth and yield [J]. Global Change Biology (3):119-128.

Blackburn G A, 1998. Spectral indices for estimating photosynthetic pigment concentrations: A test using senescent tree leaves [J]. International Journal of Remote Sensing,19 (4):657-675.

Bloom A J, Burger R, Asensio R S R, et al,2010. Carbon Dioxide Enrichment Inhibits Nitrate Assimilation in Wheat and Arabidopsis [J]. Science,328 (5980):899-903.

Bowes, 2003. Facing the Inevitable: Plants and Increasing Atmospheric CO_2 [J]. Annual Review of Plant Biology. 44 (1):309-332.

Bragazza L,2008. A climatic threshold triggers the die-off of peat mosses during an extreme heat wave [J]. Global Change Biology,14:2688-2695.

Conroy J P, Milham P J, Barlow E W R,2010. Effect of nitrogen and phosphorus availability on the growth response of Eucalyptus grandis to high CO_2 [J]. Plant,Cell and Environment,15 (7):843-847.

Drake B G, 1997. More efficient plants: A consequence of rising atmospheric CO_2? [J]. Annual Review of Plant Physiology and Plant Molecular Biology, 48 (4):609-639.

Edwards E J, Mccaffery S, Evans J R,2010. Phosphorus availability and elevated CO_2 affect biological nitrogen fixation and nutrient fluxes in a clover-dominated sward [J]. New Phytologist,169 (1):157-167.

Fang C H, Chen X S, Liu J, et al, 2008. Relation between ureide content and nitrogenase activity in soybean [J]. Journal of Northeast Agricultural University, 39 (3):9-12.

Fu L, Shao Z K,2002. Effects of CO_2 enrichment, nitrogen and soil moisture on plant C/N and C/P in spring wheat [J]. Acta Phytoecologica Sinica, 3: 295-302.

Gao J, Han X, Seneweera S, et al, 2015. Leaf photosynthesis and yield components of mung bean under fully open-air elevated [CO_2] [J]. Journal of Integrative Agriculture,3 (5):977-983.

Gifford R M,Barrett D J,Lutze J L,2000. The effects of elevated [CO_2] on the C:N and C:P mass ratios of plant tissues [J]. Plant and Soil,224 (1):1-14.

Han X, Hao X Y, Lam S K, et al, 2015. Yield and nitrogen accumulation and partitioning in winter wheat under elevated CO_2: a 3-year free-air CO_2 enrichment experiment [J]. Agriculture Ecosystems & Environment, 209: 132-137.

Hao X Y,Ji G,Xue H,et al,2014. Effects of open-air elevated atmospheric CO_2 concentration on yield quality of soybean (*Glycine max* (L.) Merr)[J]. Agriculture Ecosystems & Environment,192 (2):80-84.

Hao X Y,Han X,Lam S K,et al,2012. Effects of fully open-air [CO_2] elevation on leaf ultrastructure,photosynthesis,and yield of two soybean cultivars [J]. Photosynthetica,50 (3):362-370.

Hao X Y,Li P,Han X,et al,2016. Effects of free-air CO_2 enrichment (FACE) on N, P and K uptake of soybean in northern China [J]. Agricultural and Forest Meteorology (218):216-266.

Hatata M M, Badar R H, Ibrahim M M, et al, 2013. Respective and interactive effects of O_3 and CO_2 and drought stress on photosynthesis, stomatal conductance,antioxidative ability and yield of wheat plants [J]. Current World Environment,8:409-417.

IPCC, 2013. Summary for Policymakers. Climate Change: The Physical Science Basis. Contribution of Working Group I to the Fifth Assessment Report of the Intergovernmental Panel on Climate Change [M]. Cambridge University Press,Cambridge:United Kingdom and New York,NY,USA.

Irakli L, 2014. Hidden shift of the ionome of plants exposed to elevated CO_2 depletes minerals at the base of human nutrition. Elife. 3:e02245.

Jentsch A,Kreyling J,Elmer M,et al,2011. Climate extremes initiate ecosystem-regulating functions while maintaining productivity. Journal of Ecology, 99: 689-702.

Jiang Y L, Yao Y G, Zhang Q G, 2006. Changes of Photosynthetic Pigment Contents in Soybean under Elevated Atmospheric CO_2 Concentrations [J]. Crop Research.

John E, Major D C, Barsi A M, et al, 2007. Genetic variation and control of chloroplast pigment concentrations in *Picea rubens*, *Picea mariana* and their hybrids. I. Ambient and elevated [CO_2] environments [J]. Journal of Clinical Investigation,27 (3):353-364.

Joseph C V V, Leon H A J, 2009. Growth at elevated CO_2 delays the adverse effects of drought stress on leaf photosynthesis of the C_4 sugarcane [J]. Journal of Plant Physiology (166):107-116.

Kanowski J,2010. Effects of elevated CO_2 on the foliar chemistry of seedlings of two rainforest trees from north-east Australia: Implications for *folivorous marsupials* [J]. Austral Ecology,26 (2):165-172.

Kimball B A,2016. Crop responses to elevated CO_2 and interactions with H_2O, N, and temperature [J]. Current Opinion in Plant Biology,31:36-43.

Lam S K, Chen D, Norton R, et al, 2012a. Does phosphorus stimulate the effect of elevated [CO_2] on growth and symbiotic nitrogen fixation of grain and pasture legumes? [J]Crop and Pasture Science. 63 (1):53.

Lam S K, Chen D, Norton R, et al, 2013. The effect of elevated atmospheric carbon dioxide concentration on the contribution of residual legume and fertilizer nitrogen to a subsequent wheat crop [J]. Plant and Soil,364 (1-2): 81-91.

Lam S K, Hao X, Lin E, Han X, et al, 2012b. Effect of elevated carbon dioxide on growth and nitrogen fixation of two soybean cultivars in northern China [J]. Biology and Fertility of Soils,48 (5):603-606.

Leakey A D, Bernacchi C J, Ort D R, et al, 2006. Long-term growth of soybean at elevated [CO_2] does not cause acclimation of stomatal conductance under fully open-air conditions [J]. Plant Cell and Environment,29(9):1794-800

Leakey A D B, Xu F, Gillespie K M, et al, 2009. Genomic basis for stimulated respiration by plants growing under elevated carbon dioxide [J]. Proceedings of the National Academy of Sciences of the United States of America,106 (9):3597-3602

Li D X, Liu H L, Qiao Y Z, et al,2013a. Effects of elevated CO_2 on the growth, seed yield,and water use efficiency of soybean (*Glycine max* (L.) Merr.) under drought stress [J]. Agricultural Water Management. 129:105-112.

Li F, Kang S, Zhang J, 2003. CO_2 Enrichment on Biomass Accumulation and

Nitrogen Nutrition of Spring Wheat Under Different Soil Nitrogen and Water Status [J]. Journal of Plant Nutrition,26 (4):769-788.

Li J,Dang Q L,Man R,et al,2013b. Author's personal copy Elevated CO_2 alters N-growth relationship in spruce and causes unequal increases in N,P and K demands [J]. Forest Ecology and Management,298 (3):19-26.

Li P,Han X,Zong Y,et al,2015. Effects of free-air CO_2 enrichment (FACE) on the uptake and utilization of N,P and K in Vigna radiata [J]. Agriculture Ecosystems and Environment,202:120-125.

Long S P, Ainsworth E A, Rogers A, et al, 2004. Rising atmospheric carbon dioxide: Plants face the future [J]. Annual Review of Plant Biology, 55: 591-628.

Mastrodomenico A T,Purcell L C,King C A,2013. The response and recovery of nitrogen fixation activity in soybean to water deficit at different reproductive developmental stages [J]. Environmental and Experimental Botany,85:16-21.

Mcleod A R,Long S P,1999. Free-air Carbon Dioxide Enrichment (FACE) in Global Change Research:A Review [J]. Advances in Ecological Research,28: 1-56.

Moran J F,Becana M,Iturbe-Ormaetxe I,et al,1994. Drought induces oxidative stress in pea plants [J]. Planta,194:346-352.

Morgan P B, Bollero G A, Nelson R L, et al, 2005. Smaller than predicted increase in aboveground net primary production and yield of field grown soybean under fully open air [CO_2] elevation [J]. Global Change Biology,11: 1856-1865.

Myers D A,Thomas R B,Delucia E H,2010. Photosynthetic capacity of loblolly pine (*Pinus taeda* L.) trees during the first year of carbon dioxide enrichment in a forest ecosystem [J]. Plant,Cell and Environment,22 (5):473-481.

Niinemets U,Tenhunen J D,Canta N R,et al,2010. Interactive effects of nitrogen and phosphorus on the acclimation potential of foliage photosyntheticproperties of cork oak, *Quercus suber*, to elevated atmospheric CO_2 concentrations [J]. Global Change Biology,5 (4).

Ort D R, Ainsworth E A, Aldea M, et al, 2006. SoyFACE: The effects and interactions of elevated CO_2 and O_3 on soybean [M]. In: Nosberger J,Long

SP, Norby RJ, Stitt M, Hendrey GR, Blum H (eds) Managed ecosystems and CO_2: Case studies processes and perspectives. 71-86.

Pérez-López U, Miranda-Apodaca J, Lacuesta M, et al, 2015. Growth and nutritional quality improvement in two differently pigmented lettuce cultivars grown under elevated CO_2 and/or salinity [J]. Scientia Horticulturae, 195: 56-66.

Pang J, Zhu J G, Xie Z B, et al, 2006. A new explanation of the N concentration decrease in tissues of rice (*Oryza sativa* L.) exposed to elevated atmospheric pCO_2 [J]. Environmental and Experimental Botany, 57 (1-2): 98-105.

Pia W L, Günter N, Christof E, 2001. Elevated atmospheric CO_2 concentration favors nitrogen partitioning into roots of tobacco plants under nitrogen deficiency by decreasing nitrogen demand of the shoot [J]. Journal of Plant Nutrition. 24 (6): 835-854.

Qiao Y Z, Zhang H Z, Dong B D, et al, 2010. Effects of elevated CO_2 concentration on growth and water use efficiency of winter wheat under two soil water regimes [J]. Agricultural Water Management, 97: 1742-1748.

Razavi F, Pollet B, Steppe K, et al, 2008. Chorophyll fluorescence as a tool for evaluation of drought stress in strawberry [J]. Photosynthetica, 46: 631-633.

Reich P B, Hobbie S E, Lee T, et al, 2006. Nitrogen limitation constrains sustainability of ecosystem response to CO_2 [J]. Nature, 440 (7086): 922-925.

Rogers A, Gibon Y, Stitt M, et al, 2010. Increased C availability at elevated carbon dioxide concentration improves N assimilation in a legume [J]. Plant, Cell and Environment, 29 (8): 1651-1658.

Rohacek K, 2002. Chlorophyll Fluorescence Parameters: The Definitions, Photosynthetic Meaning, and Mutual Relationships [J]. Photosynthetica, 40 (1): 13-29.

Rosa M, Poado C, Podazz G, et al, 2009. Soluble sugars-Metabolism, sensing and abiotic stress [J]. Plant Signaling and Behavior 4: 388-393.

Roy K S, Bhattacharyya P, Neogi S, et al, 2012. Combined effect of elevated CO_2 and temperature on dry matter production, net assimilation rate, C and N allocations in tropical rice (*Oryza sativa* L.) [J]. Field Crop Research, 139 (none): 71-79.

Schwanz P, Polle A, 2001. Differential stress responses of antioxidative systems to drought in pedunculate oak (Quercus robur) and maritime pine (*Pinus pinaster*) grown under high CO_2 concentrations [J]. Journal of Experimental Botany, 52: 133-143.

Sekhar K M, Kota V R, Reddy T P, et al, 2020. Amelioration of plant responses to drought under elevated CO_2 by rejuvenating photosynthesis and nitrogen use efficiency: implications for future climate-resilient crops [J]. Photosynthesis Research.

Seneweera S, Makino A, Mae T, et al, 2005. Response of rice to p (CO_2) enrichment: the relationship between photosynthesis and nitrogen metabolism [J]. Journal of Crop Improvement, 38 (1): 289-299.

Serraj R, Sinclair T R, 2003. Evidence that carbon dioxide enrichment alleviates ureide-induced decline of nodule nitrogenase activity [J]. Annals of Botany, 1: 85-89.

Serraj R, Sinclair T R, Allen L H, 2010. Soybean nodulation and N_2 fixation response to drought under carbon dioxide enrichment [J]. Plant, Cell and Environment, 21 (5): 491-500.

Shao Y H, Wu J M, Ye J Y, et al, 2015. Frequency analysis and its spatiotemporal characteristics of precipitation extreme events in China during 1951−2010 [J]. Theoretical and Applied Climatology. 121: 775-787.

Bernstein L, 1974. Structure and Function of Plant Cells in Saline Habitats. New Trends in the Study of Salt Tolerance [J]. Translated from the Russian edition (Moscow, 1970) by A. Mercado. B. Gollek, Transl. Ed. Halsted (Wiley), New York, and Israel Program for Scientific Translations, Jerusalem, 1973. Science, 184(4141): 1067-1068.

Sushree S S, Dillip K S, Surendranath P, et al, 2015. Effect of elevated [CO_2] and nutrient management on wet and dry season rice production in subtropical India [J]. The Crop Journal, 6: 468-480.

Tausz-Posch B, Dempsey R, Norton R, et al, 2013. The effect of elevated CO_2 on photochemistry and antioxidative defence capacity in wheat depends on environmental growing conditions-A FACE study [J]. Environmental and Experimental Botany, 88: 81-92.

Vu J C V, Allen L H, 2009. Growth at elevated CO_2 delays the adverse effects of drought stress on leaf photosynthesis of the C_4 sugarcane [J]. Journal of Plant

Physiology,166 (2):107-116.

Xue X Y,Xian G L,Hai Y C,et al,2007. Effects of doubled CO_2 on AM fungi and inoculation effects on green Gram (*Phaseolus radiatus* L.) [J]. Journal of Agro-Environment Science.

Yang L,Huang J,Yang H,et al,2007a. Seasonal changes in the effects of free-air CO_2 enrichment (FACE) on nitrogen (N) uptake and utilization of rice at three levels of N fertilization [J]. Field Crop Research,100 (2-3):189-199.

Yang L X,Wang Y L,Huang J Y,et al,2007b. Seasonal changes in the effects of free-air CO_2 enrichment (FACE) on phosphorus uptake and utilization of rice at three levels of nitrogen fertilization [J]. Field Crop Research,102 (2):141-150.

Zeng Q,Liu B,Gilna B,et al,2011. Elevated CO_2 effects on nutrient competition between a C_3 crop (*Oryza sativa* L.) and a C_4 weed (*Echinochloa crusgalli* L.) [J]. Nutrient Cycling in Agroecosystems,89 (1):93-104.

Zhang X, Shang P, Qin F, et al, 2013. Chemical composition and antioxidative and anti-inflammatory properties of ten commercial mung bean samples [J]. LWT-Food Science and Technology,54 (1):171-178.

Zinta G, Abdelgawad H, Domagalska M A, et al, 2014. Physiological, biochemical, and genome-wide transcriptional analysis reveals that elevated CO_2 mitigates the impact of combined heat wave and drought stress in Arabidopsis thaliana at multiple organizational levels [J]. Global Change Biology,20:3670-3685.

Ziska L H,Bunce J A,Caulfield F A,1999. Rising atmospheric carbon dioxide and seed yield of soybean genotypes [J]. Crop Science,41:385-391.

Zong Y,Shangguan Z,2013. Short-term effects of elevated atmospheric CO_2 on root dynamics of water-stressed maize [J]. Journal of Food Agriculture and Environment,11 (1):1037-1041.

Zong Y Z,Shangguan Z P,2014. Nitrogen Deficiency Limited the Improvement of Photosynthesis in Maize by Elevated CO_2 Under Drought [J]. Journal of Integrative Agriculture,13 (1):73-81.

第三部分　大气 CO_2 浓度升高对北方主要 C_3 禾本科作物的影响

小麦(*Triticum aestivum*)作为第一大粮食作物,在世界范围内广泛种植。在未来大气 CO_2 浓度持续升高背景下,小麦生长与产量形成的适应性变化将对全球粮食生产影响较大。一项针对美国多点 FACE 平台 27 年试验结果的 meta 分析表明,小麦产量在大气 CO_2 浓度升高环境下将升高约 19%(Kimball et al.,2016)。然而,不同区域 FACE 平台小麦的增产幅度存在较大差异。在位于干旱与半干旱气候的美国亚利桑那州马里科帕 FACE 试验平台下,小麦多年平均增产 28%,正常灌溉下的提升幅度小于干旱条件(Ainsworth et al.,2005)。在同样是干旱与半干旱气候的澳大利亚 Horsham 与 Walpeup 的 Grains FACE 站,充分灌水条件下小麦产量的提高幅度高于干旱条件(Houshmandfar et al.,2016;Fitzgerald et al.,2016)。

我国是世界上小麦种植面积最大、产量最高的国家,种植面积高达 3000 万 hm^2,占世界总种植面积的 13.3%,占世界小麦总产量的 19%。所以,研究我国小麦在未来气候变化条件下的响应与适应特征对于进一步提高小麦生产能力、保障未来粮食安全具有重要意义。

第5章　大气CO_2浓度升高与不同供氮水平对小麦光合特征和氮分配的影响

5.1　引言

工业革命以来,大气CO_2浓度提升了40%,预测至21世纪末仍将不断提升(IPCC,2013)。大气CO_2浓度升高直接或间接影响了植物生长与发育(Aleminew et al.,2020)。例如,人工模拟大气CO_2浓度升高条件下,小麦的生物量增加了5%~40%(Aleminew et al.,2020;O'leary et al.,2015)。在大气CO_2浓度持续为550 $\mu mol/mol$ 时,春小麦的根系生物量与根生长速率最大增加了37%(Thilakarathne et al.,2015),冬小麦增加了75%(Ma et al.,2007)。前人研究表明,大多数作物生物量与产量在氮限制条件下可能受到抑制,进而限制了大气CO_2升高的气肥作用(Seneweera et al.,2005;Langley et al.,2010;Zong et al.,2014)。

前人对小麦的研究表明,大气CO_2升高下小麦植株的光合能力增强,进而促进了生长与产量的增长(Carlisle et al.,2012)。但是,植株对CO_2浓度升高的长期响应研究表明,大多数C_3植物的光合能力出现下调(Long et al.,2004)。原因包括:(1)氮素供应限制(Seneweera et al.,2002;Strasser et al.,2010);(2)库器官能力的欠缺(Aranjuelo et al.,2011;Ruiz-Vera et al.,2017);(3)光合基因表达与碳水化合物转运的下调(Vicente et al.,2017)。但是,Farage等(1998)的研究表明,小麦的P_n在大气CO_2浓度为500 $\mu mol/mol$ 环境中没有下调,但在700 $\mu mol/mol$ 环境中表现出明显下调(Aranjuelo et al.,2011)。

前人研究表明,小麦与其他作物叶片氮含量和光合能力都在大气CO_2浓度长期提高环境中表现出下降趋势(Bloom et al.,2010;J. Bloom et al.,2014;Zong et al.,2016)。大气CO_2浓度升高引起了氮素限制条件下的光合限制(Zong et al.,2014;Vicente et al.,2017)。作物组织中氮浓度在大气 CO_2 浓度升高环境中下降了 4.9%~19%(Ainsworth et al.,2005;Dier et al.,2018;Vicente et al.,2016)。原因包括:(1)植物生物量的增加,即"稀释作用"(Reich et al.2006);(2)由气孔导度与蒸腾降低引起的土壤——作物连续体中质流的降低(McGrath et al.,2013);(3)分配至 Rubisco 的氮含量降低(Long et al.,2004);(4)硝酸盐同化抑制(Bloom et al.,2014; Myers et al.,2014);(5)根际环境的改变与氮素可利用性的限制作用(Reich et al.,2006)。

本研究探讨了大气 CO_2 浓度升高条件下两种小麦品种叶片光合生理、生长与碳、氮浓度的适应调节特征。本研究假设:两个品种小麦的叶光合能力与器官碳氮素浓度在大气 CO_2 升高环境下的适应特征均受供氮条件影响。本研究可为探讨不同供氮水平下小麦对大气 CO_2 浓度升高环境的适应特征提供可靠依据。

5.2 试验设计与方法

本研究选取两个小麦品种 L. cv. Rac0875 和 cv. Kukri。挑选外形饱满的种子经 2% 的 NaOCl 溶液消毒 30 min,用蒸馏水反复冲洗,用 1 mM $CaCl_2$ 溶液浸种过夜,再置于 25℃恒温培养箱中萌发,每天补水,出苗大约 3 d。当种子根长至 5~6 cm 时,移入装有 1/2 Hoagland 营养液的塑料桶中培养,每桶种植 5 株幼苗,塑料桶高 12 cm、直径 5 cm,并放入人工气候室。将人工气候室 CO_2 浓度控制为 400 μmol/mol 和 700 μmol/mol 两个水平,每 3 d 更换一次 Hoagland 营养液。白天光照为 600 μmol/(m²·s),光暗周期为 12 h/12 h,昼夜温度为 25℃/20℃,昼夜空气相对湿度(RH)为 40%/55%;培养过程中剔除生长过快或过慢的植株,使待测植株长势基本一致。

供试小麦品种:Rac0875 和 Kukri。设定氮素浓度为 N1: 0.2 mM(0.1 mM NH_4NO_3),N2:0.5 mM(0.25 mM NH_4

NO_3),N3：1.0 mM(0.5 mM $NH_4 NO_3$),N4：3.0 mM(1.5 mM $NH_4 NO_3$)。于小麦生长的第4周和第6周进行气体交换参数(LI-6400型光合作用系统,LI-COR,美国)、氮素含量(凯氏定氮法,KJELTEC2300全自动定氮仪,瑞典)与碳含量(湿烧法)测定。

5.3 结果与分析

5.3.1 植物生长

在第4周和第6周分别测定了大气CO_2浓度升高与氮素供应对生物量的影响。在两种CO_2浓度下,增加氮素浓度均会增加两个小麦品种地上部与根系的生物量积累。第4周,在CO_2浓度升高环境中,不同氮素供应条件下两个小麦品种地上部生物量均没有显著增加。在第6周,几乎所有处理的地上部生物量均显著增加,Rac0875的N1处理与Kukri的N4处理除外(图5-1)。

CO_2浓度升高环境中,Rac0875的根生物量在第4周的N1和N2处理下显著增加,而N3和N4处理下显著降低,但在第6周,根生物量N1和N2处理下显著降低,而N3和N4处理下显著增加。在第4周,Kukri的根生物量在CO_2浓度升高环境中显著增加(除N1处理),在第6周也类似(除N1和N3处理)。

大气CO_2浓度升高对不同供氮条件下两个品种小麦的株高没有显著变化,高氮条件对株高具有明显促进作用。两个品种的叶面积在高氮供应下也高于低氮。同时,大气CO_2浓度升高显著增加了N1和N2处理下的Kurki,以及N4处理下的Rac0875第4周和第6周的叶面积(表5-1)。

第三部分 大气CO_2浓度升高对北方主要C_3禾本科作物的影响

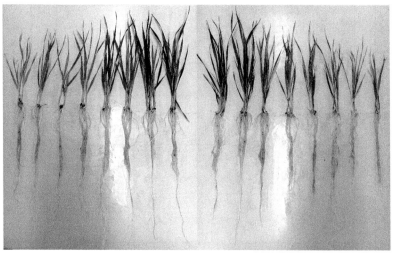

Rac0875　　　　　　　　　　Kukri

4周

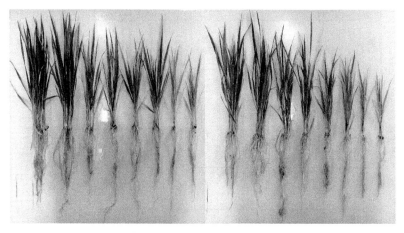

Rac0875　　　　　　　　　　Kukri

6周

图 5-1　CO_2浓度升高对不同施氮水平下小麦生长的影响

表 5-1 CO_2 浓度升高对不同施氮水平下小麦株高（cm）、干重（g）与叶面积（cm^2）的影响

品种	CO_2浓度	施氮水平	4周					6周				
			株高（cm）	干重（g）		叶面积（cm^2）		株高（cm）	干重（g）		叶面积（cm^2）	
				地上部	根系	第一叶	余叶		地上部	根系	第一叶	余叶
Rac0875	400	N1	19.15	0.17	0.14	2.85	3.44	19.15	0.35	0.25	3.99	7.98
		N2	21.53	0.27	0.17	3.8	5.62	23.45	0.62	0.38	6.87	14.96
		N3	22.95	0.33	0.23	5.77	7.21	28.18	0.87	0.46	10.24	21.95
		N4	25.23	0.5	0.3	7.38	14.07	31.5	2.16	0.92	14.4	69.08
	700	N1	19.08	0.22	0.16	3.02	3.56	19.23	0.34	0.24	3.41	6.63
		N2	21.33	0.27	0.21	3.24	4.54	22.95	0.65	0.36	6.88	12.99
		N3	23.08	0.35	0.22	5.00	7.62	28.53	1.04	0.56	11.28	24.26
		N4	25.80	0.52	0.29	7.44	15.39	32.05	2.41	1.07	14.89	72.56
Kukri	400	N1	19.35	0.18	0.16	2.08	4.05	20.03	0.29	0.27	3.2	7.89
		N2	21.23	0.23	0.17	3.53	5	24.15	0.53	0.34	5.59	12.34
		N3	23.53	0.29	0.21	4.77	8.09	31.53	0.92	0.47	9.43	24.3
		N4	27.88	0.45	0.29	6.86	15.02	35.83	1.94	0.79	12.13	58.05

第三部分 大气 CO_2 浓度升高对北方主要 C_3 禾本科作物的影响

续表

品种	CO_2浓度	施氮水平	4周					6周				
			株高(cm)	干重(g)		叶面积(cm^2)		株高(cm)	干重(g)		叶面积(cm^2)	
				地上部	根系	第一叶	余叶		地上部	根系	第一叶	余叶
Kurki	700	N1	19.63	0.2	0.15	2.55	3.78	21.53	0.36	0.32	4.07	8.57
		N2	22.6	0.29	0.22	3.87	6.08	23.93	0.57	0.34	6.29	16.71
		N3	23.93	0.34	0.23	4.54	9.04	28.95	1.13	0.61	9.28	33.71
		N4	27.3	0.47	0.32	6.86	18.74	32.88	1.95	0.77	11.5	56.41
P值		CO_2	0.44	0.08	0.02	0.69	0.08	0.36	0.11	0.05	0.36	0.28
		N	0.00	0.00	0.00	0.00	0.00	0.00	0.00	0.00	0.00	0.00
		品种	0.00	0.21	0.49	0.01	0.02	0.01	0.12	0.09	0.00	0.38
		C*N	0.91	0.98	0.21	0.39	0.14	0.66	0.74	0.27	0.87	0.62
		C*品种	0.67	0.71	0.29	0.21	0.18	0.34	0.83	0.70	0.93	0.46
		N*品种	0.19	0.86	0.83	0.22	0.47	0.73	0.1	0.00	0.00	0.00
		C*N*品种	0.49	0.81	0.38	0.78	0.63	0.30	0.77	0.34	0.15	0.60

5.3.2 光合作用

在第 6 周,N2,N3 和 N4 处理下,两个品种在大气 CO_2 浓度升高条件下均具有较高的 P_n。但在 N1 处理下,大气 CO_2 浓度升高环境中的 P_n 低于正常 CO_2 浓度环境(图 5-2)。

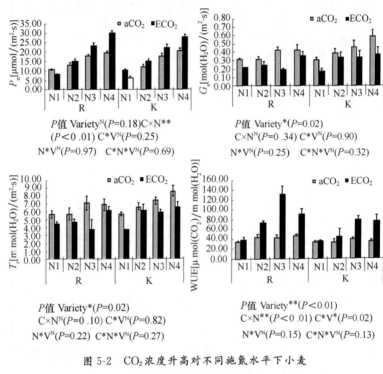

图 5-2 CO_2 浓度升高对不同施氮水平下小麦功能叶气体交换参数的影响

5.3.3 氮、碳浓度

相较于低氮条件,高氮显著提高了两种 CO_2 浓度下地上部与根系中的氮浓度。在不同供氮条件下,Rac0875 的地上部氮浓度均低于 Kukri。大气 CO_2 浓度升高下,Kukri 的地上部氮浓度降低。然

第三部分 大气 CO_2 浓度升高对北方主要 C_3 禾本科作物的影响

而,在 N4 水平下,第 4 周与第 6 周 Rac0875 的根系氮浓度显著增加,Kurki 却显著下降。

在两种 CO_2 浓度条件下,供氮水平的提高显著提升了第 4 周和第 6 周地上部与根系碳浓度。CO_2 浓度升高下,不同供氮条件下两个品种小麦的地上部碳浓度在第 4 周显著降低,但是在第 6 周显著升高。N4 处理下,两个小麦品种的根系碳浓度均在第 4 周增加(表 5-2)。

表 5-2 CO_2 浓度升高对不同施氮水平下小麦体内碳、氮浓度的影响

品种	CO_2 浓度	施氮水平	4 周				6 周			
			地上部		根系		地上部		根系	
			氮(%)	碳(%)	氮(%)	碳(%)	氮(%)	碳(%)	氮(%)	碳(%)
R	400	N1	1.17	40.92	1.15	41.69	1.32	40.45	1.2	41.69
		N2	1.4	41.07	1.38	41.39	1.4	40.81	1.36	41.84
		N3	1.81	41.6	1.49	41.3	1.77	41.66	1.5	40.94
		N4	2.58	42.23	1.91	39.52	2.26	42.27	1.6	40.62
	700	N1	1.1	40.73	1.14	40.5	1.01	40.92	1.19	41.76
		N2	1.22	40.97	1.25	40.44	1.28	41.41	1.36	42.12
		N3	1.41	41.08	1.26	41.26	1.51	41.94	1.44	41.55
		N4	2.46	41.95	1.99	40.6	2.08	42.83	1.56	40.62
K	400	N1	1.2	41.09	1.22	41.1	1.18	40.62	1.12	41.92
		N2	1.39	41.03	1.27	41.69	1.32	41.08	1.29	42.15
		N3	1.72	41.58	1.5	40.92	1.52	41.31	1.37	41.98
		N4	2.78	42.43	1.92	39.3	2.1	42.4	1.36	40.64
	700	N1	1.12	40.7	1.08	40.95	1	41.06	1.32	42.37
		N2	1.3	40.82	1.25	40.23	1.42	41.56	1.35	41.81
		N3	1.75	41.25	1.73	40.2	1.51	42.26	1.37	41.52
		N4	2.75	41.56	1.75	40.29	1.66	42.55	1.43	41.31

续表

品种	CO_2浓度	施氮水平	4周				6周			
			地上部		根系		地上部		根系	
			氮(%)	碳(%)	氮(%)	碳(%)	氮(%)	碳(%)	氮(%)	碳(%)
P值		CO_2	0.05	0.00	0.22	0.14	0.00	0.00	0.41	0.25
		N	0.00	0.00	0.00	0.00	0.00	0.00	0.00	0.00
		品种	0.07	0.89	0.63	0.22	0.01	0.32	0.02	0.03
		CO_2 * N	0.87	0.25	0.86	0.00	0.05	0.57	0.58	0.78
		CO_2 * 品种	0.21	0.26	0.51	0.88	0.33	0.83	0.09	0.56
		N * 品种	0.53	0.75	0.02	0.56	0.05	0.41	0.12	0.57
		CO_2 * N * 品种	0.57	0.37	0.02	0.45	0.11	0.05	0.81	0.10

5.4 讨论

植物生长对 CO_2 浓度升高的响应受到基因型与 CO_2 浓度的共同影响。前人研究表明,大气 CO_2 浓度升高可以提升小麦干物质积累量(Aleminew et al.,2020;Jin et al.,2019)。本研究表明,低氮供应下 Rac0875 生物量的积累与叶面积形成在大气 CO_2 浓度升高下的第 6 周显著降低。Vicente(2016)对小麦的研究也表明,大气 CO_2 浓度升高环境下开花期与灌浆期植株干物质积累在低氮条件下显著降低。也有研究表明,小麦生育前期植株生物量积累在大气 CO_2 浓度升高环境下没有响应或者呈负面响应(Aranjuelo et al.,2011;Carlisle et al.,2012)。本研究表明,低氮对光合作用的抑制降低了 Rac0875 的生物量积累。相反,大气 CO_2 浓度升高提高了高氮供应

第三部分 大气 CO_2 浓度升高对北方主要 C_3 禾本科作物的影响

下 Rac0875 干物质积累和叶面积。但第 6 周，大气 CO_2 浓度升高提高了低氮供应下 Kukri 第 6 周的干物质积累。叶面积的增加可能对 Kukri 生物量的积累起到促进作用。不同品种对大气 CO_2 浓度升高的响应策略不同体现了品种的可塑性。

本研究中，在低氮供应下，两个品种在大气 CO_2 浓度升高条件下均体现出较低的光合能力。前人对高粱（Saman et al.，2001；Zhao et al.，2010）和小麦（Vicente et al.，2015）的研究也得出了类似结论。低氮供应限制了细胞伸长和分化，进而限制了叶面积的扩展（Thilakarathne et al.，2015；Seneweera，2011）。Rac0875 的叶面积在低氮与大气 CO_2 浓度升高下降低，但 Kukri 并没有明显下降。因而，Kukri 光合能力的降低可能与光合器官氮分配降低有关。本研究表明，大气 CO_2 浓度升高与氮素供应对提高光合性能和生物量积累有积极的促进作用。

另外，本研究证明高氮供应与大气 CO_2 浓度升高显著提高了两个品种小麦的 WUE_i，这与 Novriyanti 等（2012）的结果一致。在高氮与 CO_2 浓度升高条件下，生长速率和 WUE 均显著提高。相反，不同氮素供应条件下两个品种的 G_s 与 T_r 均在大气 CO_2 浓度升高条件下显著降低，与 Katul 等（2010）的研究结果相一致。这说明大气 CO_2 浓度升高条件下 G_s 与 T_r 的降低对 WUE 起到较大的提升作用。

前人研究表明，小麦叶片氮浓度在 CO_2 浓度升高条件下显著降低（Butterly et al.，2015；Lekshmy et al.，2013）。本研究结论与上述一致，第 4 周和第 6 周 Rac0875 地上部氮浓度在大气 CO_2 浓度升高下显著降低。大多数供氮条件下，Kukri 地上部氮浓度在大气 CO_2 浓度升高下均降低（N3 水平第 4 周和 N2 水平第 6 周除外）。Bloom 等（2010）的研究表明，大气 CO_2 浓度升高限制了硝酸盐的同化，进而降低了组织氮浓度。同时，植物在大气 CO_2 浓度升高环境中积累了较多生物量，进而稀释了叶片蛋白质浓度，植物氮吸收的改变也降低了氮浓度（J. Bloom et al.，2014；Zong et al.，2016）。大气 CO_2 浓度升高对 Rac0875 与 Kukri 地上部氮浓度的降低机制需要进一步探讨。

Butterly 等(2015)的研究表明,大气 CO_2 浓度升高下整株碳浓度没有显著变化。但本研究中两个品种小麦地上部碳浓度在不同供氮水平下均显著下降。地上部碳浓度的下降可能是生物量增加的"稀释作用"引起的。但是,在第 6 周光合性能与碳积累在地上部显著升高。

5.5 结论

总之,在大气 CO_2 浓度升高条件下,高氮供应下 G_s 与 T_r 显著降低,但 P_n 与 WUE_i 显著升高。Rac0875 的地上部器官氮浓度在大气 CO_2 浓度升高下显著降低。因而,大气 CO_2 浓度升高对植株生长的促进作用受到供氮条件的限制。结果表明未来大气 CO_2 浓度升高下,小麦生产需要投入更多的氮素。

第6章 大气 CO_2 浓度升高与干旱交互对小麦光合特征和产量的影响

6.1 引言

伴随大气 CO_2 浓度的不断升高,植物细胞代谢过程与个体生长发育受到不同程度的影响(Jin et al.,2019;Kooi et al.,2016)。CO_2 浓度升高与干旱对作物生长的影响受到植物生理学家和农学家的广泛关注。通过研究 CO_2 浓度升高和水分胁迫对小麦植株光合生理、气孔特性、叶片性状、生物量积累和籽粒产量构成因素等的影响,反映其在未来气候变化条件下光合碳同化、氮素利用效率和物质生产的变化特征,为预测气候变化对旱地植物生产力的影响提供理论参考。

CO_2 浓度升高会破坏生长后期光系统Ⅱ反应中心结构,导致叶片的光合能力降低,进而影响其光合产物形成(Soba et al.,2020;Sekhar et al.,2020;Gao et al.,2015)。许育彬等(2012)研究表明,CO_2 浓度升高,抽穗期和灌浆中期 P_n 分别下降 8.7% 和 27.5%,旗叶叶绿素含量分别下降 2.6% 和 2.0%。但韩雪等(2012)研究表明,CO_2 浓度升高使小麦拔节期株高增加,单位面积穗数和穗粒数分别提高 5.3% 和 14.5%,不可孕小穗数下降 11.12%,产量增加 18.3%。CO_2 浓度升高与干旱胁迫同时发生时,CO_2 的气肥效应会受到一定程度抑制(Zong et al.,2014;李靖涛 et al.,2015),而 CO_2 浓度升高可以降低或轻微缓解干旱胁迫引起的不利作用。

气孔是植物叶片与外界进行气体交换和水分散失的主要通道,其结构与开度因环境条件的改变发生变化。前人研究表明,CO_2 浓度升高使小麦叶片气孔密度降低 14.33%,单个孔器周长和面积显著增加,叶片表皮细胞分裂增强,表皮细胞数量增加,上下表皮气孔指数下降。

前人研究表明，CO_2 浓度倍增，大豆叶片总叶面积显著增加。比叶面积(SLA)和单位生物量氮含量(N_{mass})分别表征了植物叶片对光的捕获能力和同化能力。CO_2 浓度升高，使地上部氮素含量降低，C/N 增加，植物对氮素吸收和同化的需求提高。生长在高 CO_2 浓度下叶片单位面积氮含量(N_{area})降低。

有研究表明，CO_2 浓度升高会增加植株干物质积累，具有"气肥效应"。刘月岩等研究表明，与正常 CO_2 浓度相比，CO_2 浓度升高可促进冬小麦生长，其地上生物量显著增加，正常水分和水分胁迫下分别增加 28.6% 和 18.6%。姜帅等研究表明，不考虑水分条件影响下，CO_2 浓度升高使冬小麦整个生育期株高、地上部生物量较正常 CO_2 浓度显著增加。水分胁迫降低植株叶片光合及生物量，任红旭和孟凡超等研究表明，无论是在高 CO_2 浓度还是在正常 CO_2 浓度下，水分胁迫均会降低植株地上部生物量，但高 CO_2 浓度下受旱植株地上部生物量降低幅度显著低于正常 CO_2 浓度下受旱植株。可见 CO_2 浓度升高可缓解轻度干旱胁迫的负效应。

6.2 试验设计与方法

本试验于 2015—2016 年在山西农业大学人工模拟控制 CO_2 浓度的开顶式气室(OTC)进行。生长室长 6 m，宽 4 m，高 3 m，顶开口呈八角形。供试品种为抗旱性较强的长旱 58。OTC 生长室内地面中部安装带小管辐射状的 CO_2 发散器，地面埋有与控制室 CO_2 钢瓶相连的通 CO_2 气体的管道，各生长室安装 CO_2 浓度传感器及水分测定器，可随时监测室内 CO_2 浓度，高 CO_2 浓度室内 CO_2 气体由减压阀控制的压缩 CO_2 钢瓶供给。处理期间的 CO_2 浓度由钢瓶流出，进出温室中由通气小管散开，风扇运行，保证室内 CO_2 浓度均匀，设置高 CO_2 浓度处理的目标浓度，检测仪连续监测，当室内 CO_2 浓度低于目标浓度时，自动将 CO_2 注入室内，使其浓度保持在(560±20) $\mu mol/mol$，通气时间 08:00—18:00，10 min 记录一次数据，正常

第三部分　大气 CO_2 浓度升高对北方主要 C_3 禾本科作物的影响

CO_2 浓度室内只通空气,不通入 CO_2。

本试验设置 2 个 CO_2 浓度:正常 CO_2 浓度(380 $\mu mol/mol$)和高 CO_2 浓度(560 $\mu mol/mol$);2 个水分处理:模拟根际干旱胁迫设正常水分(W:田间持水量 80%~100%)和水分胁迫(D:田间持水量 45%~55%)两个水平,重复 6 次(表 6-1)。供试小麦品种:长旱 58。采用盆栽试验,小麦播种于直径 32 cm、高 31 cm 的塑料桶中,沿桶一侧装导水管到桶底,桶底部装石子约 5 kg,上层装土,土深 26 cm,重约 13 kg。每桶种 12 穴,每穴播 8~9 粒种子,长出后每穴留苗 4~5 株。于小麦返青期进行水分处理;于小麦生长的关键时期进行气体交换参数(LI-6400 型光合作用系统,LI-COR,美国)、气孔形态(指甲油法)、氮素含量(凯氏定氮法,KJELTEC2300 全自动定氮仪,瑞典)和叶面积等光合特征与叶片性状测定,于成熟期测定籽粒产量。

表 6-1　CO_2 浓度和干旱双因素试验设计

处理	描述
$C_{380}*W$	CO_2 浓度约为 380 $\mu mol/mol$ 田间持水量 80%~100%
$C_{380}*D$	CO_2 浓度约为 380 $\mu mol/mol$ 田间持水量 45%~55%
$C_{560}*W$	CO_2 浓度约为 560 $\mu mol/mol$ 田间持水量 80%~100%
$C_{560}*D$	CO_2 浓度约为 560 $\mu mol/mol$ 田间持水量 45%~55%

6.3　结果与分析

6.3.1　叶片气孔器结构特征

CO_2 浓度升高使正常水分下小麦叶片上、下表皮气孔器周长分别显著降低 4.87%、1.89%,上表皮气孔器面积显著降低 7.05%,水分胁迫下无明显变化。两种水分条件与两种 CO_2 浓度下下表皮气孔器周长和单个气孔器面积均无明显变化。如图 6-1(a)、图 6-2 所示，CO_2 浓度升高使正常水分下小麦叶片上、下表皮气孔密度分别显著降低了 18.51%、31.41%，气孔指数分别显著降低 23%、32.67%，水分胁迫下气孔密度和气孔指数均没有显著变化。

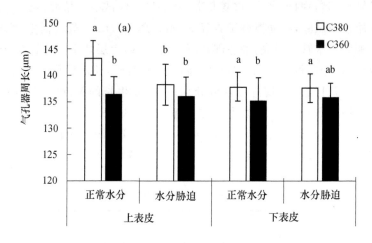

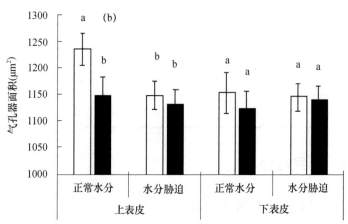

图 6-1 CO_2 浓度升高与干旱对小麦气孔结构的影响

第三部分 大气 CO_2 浓度升高对北方主要 C_3 禾本科作物的影响

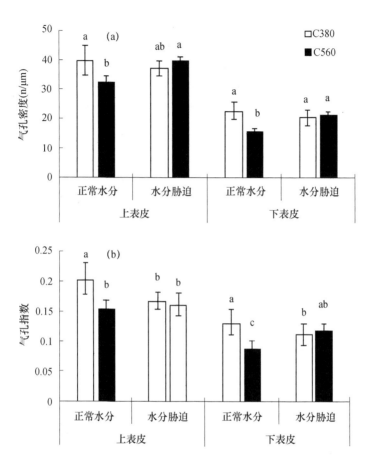

图 6-2 CO_2 浓度升高与干旱对小麦气孔密度和气孔指数的影响

6.3.2 CO_2 浓度升高与干旱下叶片经济性状的变化

如表 6-2 所示,CO_2 浓度升高下正常供水的小麦叶片叶面积(L_{area})、比叶面积(SLA)和光合氮利用率(PNUE)分别显著升高 125.05%、170.39% 和 189.34%,N_{area} 显著降低 63.74%。说明 CO_2 浓度升高虽然降低了 N_{area},但促进叶片扩大与干物质积累,提高了 PNUE。两种 CO_2 浓度下,水分胁迫使叶片 L_{area}、N_{mass} 和 N_{area} 均显著降低。在 CO_2 浓度升高环境中,水分胁迫下 NUE 没有发生明显变化;正常 CO_2 浓度环境中,水分胁迫的小麦叶片 NUE 显著降低。

表 6-2 CO_2 浓度升高与干旱对小麦叶片经济性状的影响

处理	L_{area} (cm^2)	SLA (cm^2/g)	N_{mass} (mg/g)	N_{area} (mg/cm^2)	NUE (g/mg)	PNUE [$CO_2\mu$mol/ (mg·s)]
$C_3$80 * W	4.19± 1.12 b	129.90± 97.73 b	2.40± 0.06 a	0.18± 0.00 a	0.22± 0.05 a	115.48± 3.15 d
$C_3$80 * D	2.53± 0.54 c	212.66± 48.65 b	1.71± 0.01 b	0.08± 0.00 b	0.18± 0.03 b	241.82± 2.43 b
C560 * W	9.58± 1.29 a	351.24± 53.15 a	2.31± 0.05 a	0.06± 0.00 c	0.19± 0.04 ab	334.14± 7.51 a
C560 * D	2.49± 0.50 c	204.37± 43.49 b	1.75± 0.04 b	0.08± 0.00 b	0.18± 0.04 b	178.80± 4.28 c

注:表中 L_{area} 为叶面积;SLA 为比叶面积;N_{mass} 为单位生物量氮含量;N_{area} 为单位叶面积氮含量;NUE 为氮利用效率;PNUE 为光合氮利用效率。

6.3.3 CO_2 浓度升高与干旱下小麦籽粒产量构成因素的变化特征

如表 6-3 所示,CO_2 浓度升高使正常水分下小麦茎粗、穗重、籽粒数、穗粒重和可孕小穗数分别显著增加了 13.33%、22.72%、19.8%、25.39% 和 10.76%,使水分胁迫下的小麦茎粗、穗重、籽粒数和可孕小穗数分别显著增加 21.6%、32.5%、34.71% 和 27.14%。水分胁迫下增幅高于正常水分。高浓度 CO_2 可以适当缓解水分胁迫对产量形成产生的不利影响。

表 6-3 CO_2 浓度升高与干旱对小麦籽粒产量构成因素的影响

处理	茎粗 (cm)	穗重 (g)	籽粒数 (个)	穗粒重 (g)	可孕小穗数 (个)	不可孕小穗数 (个)
$C_3$80 * W	1.80± 0.16 b	0.88± 0.27 a	20.20± 3.27 b	0.63± 0.22 b	6.50± 1.28 b	2.85± 0.49 b

第三部分　大气 CO_2 浓度升高对北方主要 C_3 禾本科作物的影响

续表

处理	茎粗 (cm)	穗重 (g)	籽粒数 (个)	穗粒重 (g)	可孕小穗数 (个)	不可孕 小穗数 (个)
C380*D	1.25± 0.11 d	0.40± 0.11 d	9.65± 3.23 d	0.31± 0.09 c	3.50± 0.76 d	3.35± 0.88 a
C560*W	2.04± 0.19 a	1.08± 0.23 a	24.20± 4.53 a	0.79± 0.18 a	7.20± 1.36 a	2.45± 0.60 b
C560*D	1.52± 0.13 c	0.53± 0.11 c	13.00± 3.43 c	0.40± 0.08 c	4.45± 0.76 c	2.65± 0.59 b

6.4　讨论

前人研究表明，CO_2 浓度升高通过促进小麦叶片伸长而增加 L_{area}（许育彬等，2012）。N_{area} 是植物光合能力的关键限制因子之一（李轩然等，2007），较高的 N_{area} 有利于提高叶片光合能力（尹丽等，2011；许振柱等，2007）。本研究结果表明，CO_2 浓度升高虽然降低了 N_{area}，但促进了叶片扩大与干物质积累（L_{area} 与 SLA），提高了 PNUE。在 CO_2 浓度升高环境中，水分胁迫的小麦叶片 NUE 显著降低，而在正常 CO_2 浓度环境中没有明显变化。

CO_2 浓度升高对小麦生物量积累的影响因时期不同而异。不同环境下的 FACE 平台试验结果表明，大气 CO_2 浓度升高有助于提高小麦地上部干物质积累及花后碳氮转运（Aleminew et al.，2020；Dier et al.，2018；Oleary et al.，2015）。不少研究表明，水分胁迫降低了 CO_2 浓度升高对小麦生物量积累的促进作用（Sekhar et al.，2020；Wullschleger et al.，2010；Li et al.，2020）。本研究中，大气 CO_2 浓度升高显著增加了开花期与灌浆期的叶干重。然而，茎干重与穗干重的积累仅在水分胁迫下受到抑制，并不受大气 CO_2 浓度的影响。

前人研究表明，大气 CO_2 浓度升高主要通过提高小麦穗数或穗粒数提高产量(Aleminew et al.,2020;Jin et al.,2019)。O'leary 等(2015)与许育彬等(2012)在 FACE 平台的试验结果表明,大气 CO_2 浓度升高下单位面积小麦分蘖数与分蘖成穗率均显著增加。韩雪等(2012)通过多年试验表明,CO_2 浓度升高下冬小麦产量增加主要是由于籽粒数增多而实现的,同时不孕小穗数可减少 17.34%～26.02%。对豆类的研究表明,高浓度 CO_2 可提高千粒重与荚数,进而增加产量(Li et al.,2013)。本研究结果表明,发生水分胁迫时,两种 CO_2 浓度下小麦茎粗、穗重、籽粒数、穗粒重、可孕小穗数均显著降低。但在高 CO_2 浓度环境中,水分胁迫引起的产量构成的降低幅度小于正常 CO_2 浓度。说明提高 CO_2 浓度对水分胁迫造成产量构成因素的降低有缓解作用。

6.5 结论

正常水分条件下,CO_2 浓度升高通过降低上表皮气孔密度、气孔器周长、单个气孔器面积与瞬时 G_s,使 T_r 维持在较低水平;但在水分胁迫时,高浓度 CO_2 下上表皮气孔密度、气孔器周长及面积较正常 CO_2 浓度下小麦叶片维持不变。正常水分条件下,CO_2 浓度升高并未改变叶 N_{mass},但却使叶面积与 SLA 升高,从而降低了 N_{area},提高了 PNUE;水分胁迫时,高 CO_2 浓度下 L_{area}、SLA、N_{area}、N_{mass} 均较正常 CO_2 浓度未显著降低。水分胁迫降低了小麦生物量积累和籽粒产量形成,但 CO_2 浓度升高可缓解水分胁迫带来的不利影响。

参考文献

韩雪,郝兴宇,王贺然,等,2012.FACE 条件下冬小麦生长特征及产量构成的影响[J].中国农学通报,28(36):154-159.

李靖涛,居辉,王宏富,2015.不同水分条件下 CO_2 浓度升高对冬小麦碳氮转运的影响[J].中国生态农业学报(8):954-963.

李轩然,刘琪璟,蔡哲,2007.千烟洲针叶林的比叶面积及叶面积指数[J].植物生态学报(1):93-101.

第三部分　大气 CO_2 浓度升高对北方主要 C_3 禾本科作物的影响

许育彬,沈玉芳,李世清,2012. 分期施氮与 CO_2 浓度升高对小麦光合及其物质积累和产量的互作效应[J]. 西北植物学报,32(5):1007-1012.

许振柱,周广胜,2007. 全球变化下植物的碳氮关系及其环境调节研究进展——从分子到生态系统[J]. 植物生态学报,31(4):738-747.

尹丽,胡庭兴,刘永安,2011. 施氮量对麻疯树幼苗生长及叶片光合特性的影响[J]. 生态学报,31(17):4977-4984.

Ainsworth E A, Long S P, 2005. What have we learned from 15 years of free-air CO_2 enrichment (FACE)? A meta-analytic review of the responses of photosynthesis, canopy properties and plant production to rising CO_2[J]. New Phytologist, 165 (2):351-372.

Aleminew A, Abera M, 2020. Effect of Climate Change on the Production and Productivity of Wheat Crop in the Highlands of Ethiopia [J]. A Review. 41 (1):34-42

Aranjuelo I, Cabrera-Bosquet L, Morcuende R, et al, 2011. Does ear C sink strength contribute to overcoming photosynthetic acclimation of wheat plants exposed to elevated CO_2? [J]. Journal of Experimental Botany, 62 (11):3957-3969.

Bloom A J, Burger R, Asensio R S R, et al, 2010. Carbon Dioxide Enrichment Inhibits Nitrate Assimilation in Wheat and Arabidopsis [J]. Science, 328 (5980):899-903.

Butterly C R, Armstrong R, Chen D L, et al, 2015. Carbon and nitrogen partitioning of wheat and field pea grown with two nitrogen levels under elevated CO_2[J]. Plant and Soil, 391 (1-2):367-382.

Carlisle E, Myers S, Raboy V, et al, 2012. The effects of inorganic nitrogen form and CO_2 concentration on wheat yield and nutrient accumulation and distribution [J]. Frontiers in Plant Science, 3 (4):195.

Dier M, Meinen R, Erbs M, et al, 2018. Effects of free air carbon dioxide enrichment (FACE) on nitrogen assimilation and growth of winter wheat under nitrate and ammonium fertilization [J]. Global Change Biology, 24 (1): e40-e54.

Farage P K, McKee I F, Long S P, 1998. Does a Low Nitrogen Supply Necessarily Lead to Acclimation of Photosynthesis to Elevated CO_2? [J]. Plant Physiology, 118 (2):573.

Fitzgerald G J, Tausz M, OLeary G, et al, 2016. Elevated atmospheric [CO_2] can dramatically increase wheat yields in semi-arid environments and buffer against heat waves [J]. Global Change Biology,22:2269-2284.

Gao J, Han X, Seneweera S, et al, 2015. Leaf photosynthesis and yield components of mung bean under fully open-air elevated [CO_2] [J]. Journal of Integrative Agriculture,3(5):977-983.

Houshmandfar A, Fitzgerald G J, Macabuhay A A, et al, 2016. Trade-offs between water-use related traits, yield components and mineral nutrition of wheat under free-air CO_2 enrichment (FACE) [J]. European Journal of Agronomy,76:66-74.

IPCC,2013. Summary for Policymakers. Climate Change: The Physical Science Basis. Contribution of Working Group I to the Fifth Assessment Report of the Intergovernmental Panel on Climate Change [M]. Cambridge University Press,Cambridge,United Kingdom and New York,NY,USA.

Bloom A,Burger M,Kimball B,et al,2014. Nitrate assimilation is inhibited by elevated CO_2 in field-grown wheat [J]. Nature Climate Change, 4 (6): 477-480.

Jin J, Armstrong R, Tang C, 2019. Impact of elevated CO_2 on grain nutrient concentration varies with crops and soils-A long-term FACE study [J]. Science of The Total Environment,651:2641-2647.

Katul G, Manzoni S, Palmroth S, 2010. A stomatal optimization theory to describe the effects of atmospheric CO_2 on leaf photosynthesis and transpiration [J]. Annals of Botany,105 (3):431-442.

Kimball,Bruce A, 2016. Crop responses to elevated CO_2 and interactions with H_2O,N,and temperature [J]. Current Opinion in Plant Biology,31:36-43.

Kooi C J, Reich M, Löw M, et al, 2016. Growth and yield stimulation under elevated CO_2 and drought: a meta-analysis on crops [J]. Environmental and Experimental Botany,122:150-157.

Langley J A, Megonigal J P, 2010. Ecosystem response to elevated CO_2 levels limited by nitrogen-induced plant species shift [J]. Nature Genetics, 466 (7302):96-99.

Lekshmy S,Jain V, Khetarpal S,et al,2013. Inhibition of nitrate uptake and assimilation in wheat seedlings grown under elevated CO_2[J]. Indian Journal of Plant Physiology,18 (1):23-29.

Li D X, Liu H L, Qiao Y Z, et al, 2013. Effects of elevated CO_2 on the growth, seed yield, and water use efficiency of soybean (*Glycine max* (L.) Merr.) under drought stress [J]. Agricultural water management, 129:105-112.

Li Y, Li S, He X, et al, 2020. CO_2 enrichment enhanced drought resistance by regulating growth, hydraulic conductivity and phytohormone contents in the root of cucumber seedlings [J]. Plant Physiology and Biochemistry, 152: 62-71.

Long S P, Ainsworth E A, Rogers A, et al, 2004. Rising atmospheric carbon dioxide: Plants face the future [J]. Annual Review of Plant Biology, 55: 591-628.

Ma L, Zhu J G, Xie Z B, 2007. Responses of rice and winter wheat to free-air CO_2 enrichment (China FACE) at rice/wheat rotation system [J]. Plant and Soil, 294 (1-2):137-146.

McGrath J M, Lobell D B, 2013. Reduction of transpiration and altered nutrient allocation contribute to nutrient decline of crops grown in elevated CO_2 concentrations [J]. Plant, Cell and Environment, 36 (3):697-705.

Myers S S, Zanobetti A, Kloog I, et al, 2014. Increasing CO_2 threatens human nutrition [J]. Nature, 510 (7503):139-142.

Novriyanti E, Watanabe M, Kitao M, et al, 2012. High nitrogen and elevated [CO_2] effects on the growth, defense and photosynthetic performance of two eucalypt species [J]. Environmental Pollution, 170:124-130.

Oleary G J, B. Christy J, et al, 2015. Response of wheat growth, grain yield and water use to elevated CO_2 under a free-air CO_2 enrichment (FACE) experiment and modeling in a semi-arid environment [J]. Global Change Biology, 7 (21):2670-2686.

Reich P B, Hobbie S E, Lee T, et al, 2006. Nitrogen limitation constrains sustainability of ecosystem response to CO_2 [J]. Nature. 440 (7086):922-925.

Ruiz-Vera U M, De Souza A P, Long S P, et al, 2017. The Role of Sink Strength and Nitrogen Availability in the Down-Regulation of Photosynthetic Capacity in Field-Grown Nicotiana tabacum L. at Elevated CO_2 Concentration [J]. Front Plant Science, 8:998.

Saman S, Oula G, Jann P C, 2001. Root and shoot factors contribute to the effect of drought on photosynthesis and growth of the C_4 grass Panicum coloratum at elevated CO_2 partial pressures [J]. Australian Journal of Plant Physiology, 28 (6):451-460.

Sekhar K M, Kota V R, Reddy T P, et al, 2020. Amelioration of plant responses to drought under elevated CO_2 by rejuvenating photosynthesis and nitrogen use efficiency: implications for future climate-resilient crops [J]. Photosynth Research.

Seneweera S, 2011. Reduce nitrogen allocation to expanding leaf blades leads to suppression of Ribulose-1, 5-bisphosphate carboxylase/oxygenase synthesis and photosynthetic acclimation to elevated CO_2 in rice [J]. Photosynthetica, 49 (1):145-148.

Seneweera S, Makino A, Mae T, et al, 2005. Response of rice to pCO_2 enrichment: the relationship between photosynthesis and nitrogen metabolism [J]. Journal of Crop Improvement, 38 (1):289-299.

Seneweera S P, Ghannoum O, Conroy J P, et al, 2002. Changes in source-sink relations during development influence photosynthetic acclimation of rice to free air CO_2 enrichment (FACE) [J]. Functional Plant Biology, 29:945-953.

Soba D, Shu T, Runion G B, et al, 2020. Effects of elevated [CO_2] on photosynthesis and seed yield parameters in two soybean genotypes with contrasting water use efficiency [J]. Environmental and Experimental Botany, 178:104154.

Strasser R J, Tsimilli-Michael M, Qiang S, et al, 2010. Simultaneous in vivo recording of prompt and delayed fluorescence and 820-nm reflection changes during drying and after rehydration of the resurrection plant Haberlea rhodopensis [J]. Biochimica et Biophysica Acta (BBA)-Bioenergetics, 1997 (6):1313-1326.

Thilakarathne C L, S. Tausz-Posch K, Cane R M, et al, 2015. Intraspecific variation in leaf growth of wheat (*Triticum aestivum*) under Australian Grain Free Air CO_2 Enrichment (AGFACE): Is it regulated through carbon and/or nitrogen supply? [J] Functional Plant Biology, 42 (3):299-308.

Vicente R, Pérez P, Martínez-Carrasco R, et al, 2016. Metabolic and Transcriptional Analysis of Durum Wheat Responses to Elevated CO_2 at Low and High Nitrate Supply [J]. Plant and Cell Physiology, 57 (10):2133-2146.

Vicente R, Pérez P, Martínez-Carrasco R, et al, 2017. Improved responses to elevated CO_2 in durum wheat at a low nitrate supply associated with the upregulation of photosynthetic genes and the activation of nitrate assimilation

[J]. Plant Science,260:119-128.

Vicente R, Perez P, Martinez-Carrasco R, et al, 2015. Quantitative RT-PCR platform to measure transcript levels of C and N metabolism-related genes in durum wheat:Transcript profiles in elevated [CO_2] and high temperature at different nitrogen supplies [J]. Plant and Cell Physiology,56 (8):1556-1573.

Wullschleger S D, Tschaplinski T J, Norby R J, 2010. Plant water relations at elevated CO_2-implications for water-limited environments [J]. Plant,Cell and Environment,25 (2):319-331.

Zhao X,Mao Z,Xu J,2010. Gas exchange,chlorophyll and growth responses of Betula Platyphylla seedlings to elevated CO_2 and nitrogen [J]. International Journal of Biological,2 (1):143-149.

Zong Y, Wang W, Xue Q, et al, 2014. Interactive effects of elevated CO_2 and drought on photosynthetic capacity and PS Ⅱ performance in maize [J]. Photosynthetica,52:63-70.

Zong Y Z,Shangguan Z P,2014. Nitrogen Deficiency Limited the Improvement of Photosynthesis in Maize by Elevated CO_2 Under Drought [J]. Journal of Integrative Agriculture,13 (1):73-81.

Zong Y Z, Shangguan Z P, 2016. Increased sink capacity enhances C and N assimilation under drought and elevated CO_2 conditions in maize [J]. Journal of Integrative Agriculture,15 (012):2775-2785.

第四部分 大气 CO_2 浓度升高对北方主要 C_4 作物的影响

与 C_3 植物相比，C_4 植物叶片的光合作用具有 CO_2 浓缩机制，在维管束鞘细胞叶绿体中含有 1,5-二磷酸羧化/加氧酶，进而增强了光合羧化效率，抑制了加氧反应，减少了光呼吸（Ghannoum et al., 2011; Liu et al., 2013）。在正常大气 CO_2 浓度下，C_4 植物具有比 C_3 植物更高的光合效率（Liu et al., 2013）。所以有研究认为，在大气 CO_2 浓度长期升高环境中，C_4 植物的光合性能提高幅度较小，但仍存在较大争议。

研究表明，在正常供水条件下，玉米（Leakey et al. 2006）与谷子（郝兴宇等，2010）的光合速率在大气 CO_2 浓度升高环境中均没有显著变化，高粱产量甚至呈下降趋势（Kimball et al., 2016）。然而，美国亚利桑那州马里科帕（Arizona, USA）与伊利诺伊州（Illinois, USA）的两个 FACE 平台的多年研究数据表明，C_4 作物的生物量积累在大气 CO_2 浓度持续升高的环境中平均增长了 15%～20%（Ainsworth et al., 2005; Sekhar et al., 2020）。C_4 植物虽然仅占陆地生态系统 1% 的种类，但是其对初级生产力的贡献率却高达 21%，对全球农业籽粒产量的贡献率超过 30%（Steffen et al., 2003），其响应规律是预测未来气候变化下农业生态系统响应与适应策略的重要组成部分。

在与不同非生物胁迫的交互作用下，CO_2 的气肥作用在不同区域间存在较大差异。在我国北方，干旱与氮缺乏等是农业生态系统生产力提升的主要限制因子，C_4 作物生长与产量形成在气候变化与多重因子交互下的响应、适应规律亟待进一步阐明，对于揭示未来作物生产与粮食安全具有重要意义。

第 7 章 大气 CO_2 浓度升高对氮胁迫下受旱玉米光合特征的影响

7.1 引言

土壤干旱与氮素缺乏是全球大部分地区限制作物生长的主要因素(Mu et al.,2007;Farooq et al.,2009;Chaves et al.,2008)。虽然,氮肥的施用对缓解土壤氮素缺乏起到重要作用,但是仍然有许多发展中国家由于经济与生态等因素的制约,使氮肥的施用达不到作物生长需求(Galloway et al.,2008)。大气中 CO_2 浓度升高增强了植物生长对氮素的需求,加重了缺氮地区氮肥对作物生长的制约,进而限制了高浓度 CO_2 的气肥作用(Luo et al.,2004;Reich et al.,2006)。在氮素缺乏地区,高浓度 CO_2 常给植物生长带来负面影响(Reich et al.,2006)。

目前,已有许多研究集中于高浓度 CO_2 条件下 C_3 植物的生长调节机制(Seneweera et al.,2011;Robredo et al.,2010;Prieto et al.,2012;Long et al.,2004)。高浓度 CO_2 使植株叶片氮素含量减少,进而引起 C_3 植物的光合能力下调(Stitt,1991;Zhou et al.,2009)。在 C_4 植物中,由于叶片可以利用少量的 Rubisco 达到较高的光合效率,所以叶片氮含量的减少对其光合能力的影响较小(Furbank et al.,1996;Ghannoum et al.,2000b)。同时,由于 C_4 植物在正常 CO_2 浓度下光合速率已接近饱和,理论上 C_4 植物对大气 CO_2 浓度升高敏感性较弱,所以氮素缺乏对其在未来气候变化条件下光合能力的限制作用在目前研究较少。

在水分胁迫条件下,植株叶片的气孔限制与非气孔限制是影响光合作用调节与适应的主要因素(Ghannoum,2009;Farquhar et al.,1982;Long et al.,2003)。气孔限制常常决定了叶片胞间 CO_2 的浓度,其数值可以通过分析 A/C_i 曲线得出,即在饱和光强下测定叶片最大光合速率,此时空气中 CO_2 被认为最大限度地进入叶片(Farquhar et al.,1982)。非气孔限制包括叶片生物化学与结构特性,其数值可以通过 PEPC 羧化能力(V_{pmax})和 Rubisco 羧化能力(V_{max})体现(von Caemmerer et al.,1999)。

在 C_4 植物中,CO_2 浓度升高被证明可以减轻干旱胁迫对光合能力的限制(Leakey,2009;Ghannoum et al.,2000b;Albert et al.,2011;Allen et al.,2011),而氮素亏缺则通过减少相关酶活性加深干旱胁迫(Ghannoum et al.,1998;Galloway et al.,2008;Ghannoum et al.,2005)。但是在未来多种环境因子共同变化的条件下,氮素胁迫对高浓度 CO_2 缓解干旱对光合气孔限制和非气孔限制的影响还需要进一步研究。本章内容以 3 个假设为基础:(1)高浓度 CO_2 可以缓解干旱胁迫引起的气孔限制,而对非气孔限制缓解甚微;(2)氮素缺乏可以加强干旱胁迫下的非气孔限制,所以减少高浓度 CO_2 对干旱的缓解作用;(3)在高浓度 CO_2 与氮素缺乏条件下,玉米可以通过叶片中较少的氮含量维持较高的光合效率。

7.2 试验设计与方法

本研究挑选外形饱满的玉米种子(郑单 958)经 2% 的 NaOCl 溶液消毒 30 min,用蒸馏水反复冲洗,用 0.001mol/L $CaCl_2$ 溶液浸种过夜,再置于 25℃ 恒温培养箱中萌发,每天补水,出苗大约 3 d。当种子根长至 5~6 cm 时,移入装有 1/2 Hoagland 营养液或干净石英砂的塑料桶中培养,每桶种植 6 株幼苗,塑料桶高 15 cm,直径上为 18 cm、下为 13 cm。并放入人工气候室(每个气候室控制面积 12 m^2,杭州求是人工环境有限公司,杭州)。将人工气候室 CO_2 浓度控制为 380 $\mu mol/mol$ 和 750 $\mu mol/mol$ 两个水平,每 3 d 更换一次 Hoagland 营养液。白天光照为 600 $\mu mol/(m \cdot s)$,光暗周期为 12 h/12 h,昼夜温度为 25℃/20℃,昼夜空气相对湿度(RH)为 40%/55%;培养过程中剔除生长过快或过慢的植株,使待测植株长势基本一致。

从 80％的幼苗完全展开第三片叶片开始,在营养液中加入 10％的 PEG-6000(-0.32 MPa)作为干旱处理。从干旱胁迫的第 6 d 开始到实验结束,将部分幼苗营养液中的 KNO_3 和 $Ca(NO_3)_2$ 用 KCL 和 $CaSO_4$ 替代,即以无氮素的营养液作为氮素限制处理。试验共进行 30 d,期间每 3 d 更换一次 Hoagland 营养液,并将其 pH 值控制在 5.8~6.2。$A-C_i$ 曲线(LI-6400 型光合作用系统,LI-COR,美国)于干旱胁迫后第 14 d 进行测定,生物量与氮素含量(凯氏定氮法,KJELTEC2300 全自动定氮仪,瑞典)测定于干旱胁迫后第 0~480 h,每 72 h 进行一次。

本研究中每次取样均设有 5 个重复,试验处理如下所示:
Control:380 μmol/mol CO_2+正常水分(对照)
C-N:380 μmol/mol CO_2+正常水分+氮限制
D:380 μmol/mol CO_2+10％ PEG(干旱)
D-N:380 μmol/mol CO_2+10％ PEG(Drought)+氮限制
C(750):750 μmol/mol CO_2+正常水分
C(750)-N:750 μmol/mol CO_2+正常水分+氮限制
C(750)+D:750 μmol/mol CO_2+10％ PEG(干旱)
C(750)+D-N:750 μmol/mol CO_2+10％ PEG(干旱)+氮限制

7.3 结果与分析

7.3.1 高浓度 CO_2 与氮胁迫对生物量积累的影响

如图 7-1 所示,水分胁迫后 144 h 内,各处理之间生物量积累没有显著差异。水分胁迫 144 h 至试验结束,正常供氮使植株生物量积累高于氮素限制处理。在相同水分条件下,CO_2 倍增使植株生物量积累显著升高,在氮素限制处理中也是如此。在相同 CO_2 浓度条件下,充足水分植株的生物量积累显著高于水分胁迫的植株。

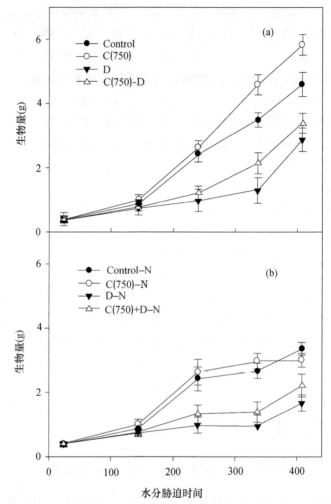

图 7-1 正常氮素(a)和氮素限制(b)条件下水分胁迫后 408 h 内玉米幼苗整株生物量的变化

7.3.2 叶片气孔与非气孔限制对氮胁迫的响应

图 7-2 表示了水分与氮素限制条件下,CO_2 倍增对玉米幼苗叶片非气孔限制的影响。不论是水分胁迫还是氮素限制,CO_2 倍增处理

第四部分 大气 CO_2 浓度升高对北方主要 C_4 作物的影响

均增强了植株的 V_{pmax} 值。氮素限制条件下,水分胁迫降低了玉米幼苗的 V_{pmax} 值,而 CO_2 倍增处理增强了水分胁迫植株的 V_{pmax} 和 V_{max} 值。

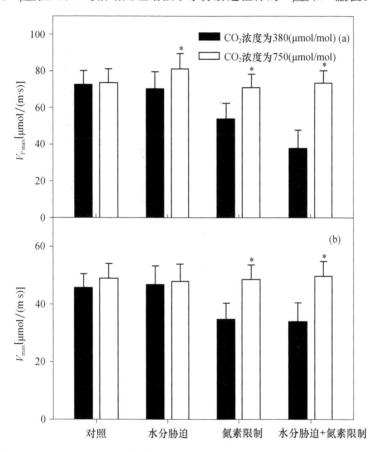

图 7-2 水分胁迫、氮素限制与 CO_2 倍增水平下玉米幼苗光合羧化效率的变化
(a) PEPC 最大羧化效率($V_{p\,max}$);(b) A/C_i 曲线初始斜率(V_{max})

在正常大气 CO_2 浓度下,氮素限制加剧了水分胁迫对植株的气孔限制,CO_2 倍增显著减轻了这种限制(图 7-3)。在正常供水且氮素限制的处理中,CO_2 倍增不能减轻植株的气孔限制值,A_{sat} 较正常 CO_2 没有显著提高。在水分胁迫条件下,正常大气 CO_2 浓度下 A_{sat}

显著低于 CO_2 倍增处理(图 7-4)。

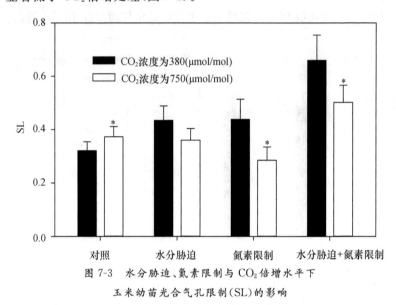

图 7-3　水分胁迫、氮素限制与 CO_2 倍增水平下玉米幼苗光合气孔限制(SL)的影响

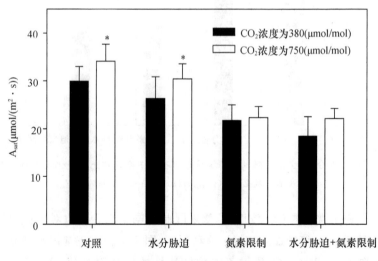

图 7-4　水分胁迫、氮素限制与 CO_2 倍增对玉米幼苗最大光合速率(A_{sat})的影响

7.3.3 叶片氮素利用效率对氮胁迫的响应

如图 7-5 所示,在正常大气 CO_2 浓度下,氮素限制的植株光合氮素利用效率(PNUE)较其余处理最低,而水分与氮素共同限制的植株 PNUE 达到所有处理中最高值。CO_2 浓度倍增使水分胁迫与氮素限制植株的 PNUE 均得到显著提高。在所有限制条件下,CO_2 倍增均显著降低了单位叶面积氮含量(N_{area})。在两种 CO_2 浓度下,N_{area} 均在水分与氮素共同胁迫的植株中达到最小,在没有胁迫的植株中达到最大。植株整株氮素利用效率(NUE)在 CO_2 倍增条件下氮素与水分共同胁迫的处理中达到所有处理中最小值。

图 7-5 表示了叶片氮含量在植株生长过程中的增长情况。在氮素充足条件下,两种 CO_2 浓度均使正常水分的植株叶片氮含量迅速增长。在氮素限制处理中,CO_2 浓度倍增较正常 CO_2 浓度显著增加了生长后期水分胁迫植株叶氮含量。

图 7-6 表示了光合羧化能力与叶片氮含量之间的相关关系,其中 $V_{pmax(CO_2=750)}/V_{pmax(CO_2=380)}$ 和 $V_{max(CO_2=750)}/V_{max(CO_2=380)}$ 均与 $N_{area(CO_2=750)}/N_{area(CO_2=380)}$ 呈显著负相关关系(V_{pmax} 中 $R^2=0.97$, $P<0.0001$;V_{max} 中 $R^2=0.65$;$P=0.016$)。$A_{sat(CO_2=750)}/A_{sat(CO_2=380)}$ 与 $N_{area(CO_2=750)}/N_{area(CO_2=380)}$ 之间相关性不显著($R^2=0.49$;$P=0.06$)。

7.4 讨论

目前,已有不少研究集中于 CO_2 浓度升高对氮素匮乏土壤中 C_3 作物生长的影响(Seneweera 等,2011;Prieto 等,2012;Ghannoum 等,1998),但对于 C_4 作物却少有研究。氮素亏缺常常引起植物光合能力的下调。尤其在土壤较为贫瘠的干旱半干旱地区,水分与氮素亏缺的共同作用会对生态系统植被生产力造成巨大影响(Ghannoum,2009;Ghannoum et al.,2011)。在大气 CO_2 浓度升高条件下,研究玉米幼苗光合特性对氮素与水分亏缺的响应策略是预测未来气候变化植被生产力的重要组成部分(图 7-6、图 7-7)。

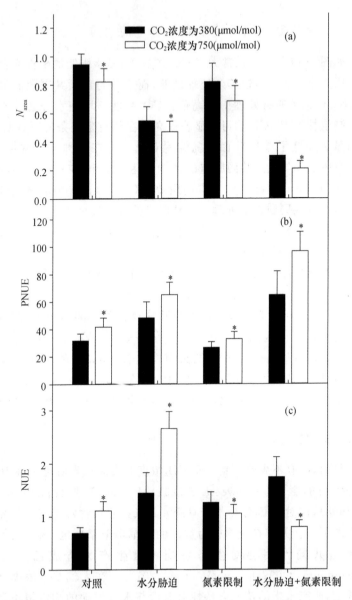

图 7-5 水分胁迫、氮素限制与 CO_2 倍增水平下玉米幼苗氮素利用效率的影响
(a)功能叶单位叶面积氮含量(N_{area});(b)光合氮利用效率(PNUE);(c)整株氮利用效率(NUE)

第四部分 大气 CO_2 浓度升高对北方主要 C_4 作物的影响

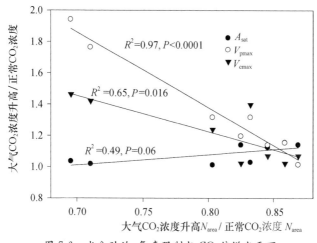

图 7-6 水分胁迫、氮素限制与 CO_2 倍增水平下玉米幼苗 A_{sat}、V_{pmax}、V_{max} 与 N_{area} 之间的关系

7.4.1 氮胁迫限制了高浓度 CO_2 对光合能力的促进作用

本研究表明，在正常大气 CO_2 条件下，水分胁迫降低了叶片 PEPC 的羧化能力（V_{pmax}），增强了气孔限制，进而抑制最大光合速率（A_{sat}）。在氮素充足条件下，CO_2 浓度倍增增加了叶片胞间 CO_2 浓度（Markelz et al.，2011），缓解了由水分胁迫引起的 PEPC 的羧化能力与气孔导度的降低，进而显著减轻了水分胁迫对气孔开闭与 V_{pmax} 的限制。在高粱的研究中，高浓度 CO_2 也被证明有缓解干旱胁迫引起的光合能力下调的作用（Vu et al.，2009）。但是，CO_2 倍增对缓解水分亏缺植株的 Rubisco 羧化能力（V_{max}）没有显著作用。

在正常 CO_2 浓度下，氮素亏缺可以加剧干旱胁迫对植株生理功能的限制作用（Ghannoum et al.，2011）。本研究表明，由水分亏缺引起的 V_{pmax} 和 V_{max} 降低及气孔限制在氮素限制的条件下显著增强，但 CO_2 浓度倍增使二者的限制作用得到缓解。然而在氮素限制条件下，CO_2 倍增对气孔限制的减轻作用并没有满足植株生长需求，其对光合能力的缓解主要通过解除非气孔限制实现。所以，氮素限制

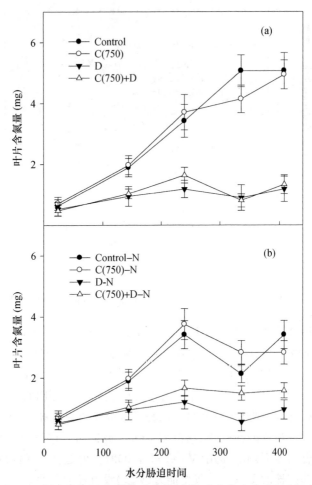

图 7-7 充足供氮(a)和氮素限制(b)条件下干旱胁迫后 408 h 内玉米植株叶片氮含量的变化

减弱了高浓度 CO_2 对受旱植株光合作用的缓解能力。在干旱半干旱地区,高浓度 CO_2 对植株光合速率的上调作用在氮素充足的生态系统中高于氮素匮乏的生态系统。

7.4.2 高浓度 CO_2 提高了氮胁迫植株的 PNUE，降低了 NUE

大气 CO_2 浓度升高可以降低氮素在植株成熟叶片中的分配(Zhou et al.,2009;Seneweera et al.,2011)。本研究进一步证明,氮素限制引起植株功能叶单位叶面积氮含量(N_{area})显著降低,CO_2 浓度倍增使这种限制加剧。但是,植株可以通过调节光合器官之间的氮素分配比例,利用较少的氮素提高光合羧化能力,增强光合氮利用效率(PNUE),进而适应高浓度 CO_2 环境(Ellsworth et al.,2004; Crous et al.,2008)。本研究通过分析 V_{pmax} 和 V_{max} 与 N_{area} 之间的相关性来阐明光合能力与氮素含量关系的变化特征。研究表明,V_{pmax} 和 V_{max} 均与 N_{area} 显著负相关,其中 CO_2 浓度倍增使水分和氮素共同亏缺的植株具有最高的羧化效率与最小的 N_{area},显著增强了叶片的 PNUE。在 FACE 实验中,草地和树木的 V_{max} 与 N_{area} 存在同样规律(Ellsworth et al.,2004),但是长期适应后,氮素在羧化作用相关酶类中的分配比率降低,导致 V_{max} 与 N_{area} 之间的斜率逐渐降低(Crous et al.,2008)。

然而,在整株水平,CO_2 浓度倍增却降低了水分胁迫条件下氮亏缺植株的 NUE。因为氮素限制条件下,CO_2 倍增使植株运输较多氮素至叶片,扩大了光合产物的消耗(Taiz et al.,2006)。所以,高浓度 CO_2 对玉米生物量积累的促进作用受到抑制,在干旱胁迫条件下这种抑制尤为明显。许多关于农作物、草种和木本植株的研究表明,氮素缺乏限制 CO_2 浓度增加对生物量的促进作用(Zanetti et al.,1996;Schneider et al.,2004;Tobita et al.,2011;Cabrera-Bosquet et al.,2009)。所以在未来气候变化条件下氮素缺乏对 CO_2 气肥作用的限制会普遍存在(Reich et al.,2006)。然而,C_4 植物生长对全球气候变化的响应仍然存在大量不确定性,大量氮素与水分水平的设计实验仍需要进一步进行。

7.5 小 结

本章阐述了氮素限制对高浓度 CO_2 缓解干旱对玉米幼苗光合

胁迫作用的影响。在氮素充足的植株中,高浓度 CO_2 通过减轻气孔限制和 V_{pmax} 限制缓解干旱引起的光合能力下调,但是对 V_{max} 限制没有减轻作用。氮素限制加重了干旱条件下光合作用的气孔限制和非气孔限制,其中非气孔限制在 CO_2 浓度倍增环境下得到一定程度的缓解,而气孔限制的缓解没有达到生长需求。CO_2 倍增使氮素亏缺植株运输至叶片的氮素增多,进而使其 PNUE 升高而 NUE 降低。所以,在氮素限制条件下,CO_2 倍增对植株干旱胁迫的缓解没有使整株氮利用效率和生物量积累增强。

第8章 大气 CO_2 浓度升高与氮胁迫对玉米幼苗干旱—复水过程中光合特性的影响

8.1 引言

大气中 CO_2 浓度与温室气体的排放引起温室效应,改变全球降雨格局(Solomon,2007),加剧了局部地区的干旱程度。在土壤贫瘠且降雨量小而集中的生态系统中,氮素缺乏会限制高浓度 CO_2 对植株生长的刺激作用(Luo et al.,2004;Reich et al.,2006)。例如,我国西北黄土高原的干旱与半干旱地区是典型的雨养农业区和生态脆弱区,年均降雨量小、雨期集中、太阳辐射强、蒸发剧烈、土壤肥力低(山仑等,2000),植被往往经历着干旱—降雨—干旱的反复过程。在未来气候变化趋势下,该类型生态系统中植物生产力及生理功能对干旱与复水循环过程的响应机制是科学家关注的热点与前沿问题,也是预测陆地生态系统对未来气候变化适应方式中不可或缺的重要组成。

在干旱的生态系统中,降雨往往使植被生产力产生波动(Reynolds et al.,2004)。Yahdjian 等(2006)在对南美洲巴塔哥尼亚大草原的研究中指出,前期干旱会抑制地上部净初级生产力,在强降雨后无法得到恢复。Xu 等(2009)也报道干旱—复水过程抑制了我国呼伦贝尔草原羊草生物量的积累。干旱胁迫易引起可溶性糖在功能叶片中积累(Dosio et al.,2011),阻碍叶面积扩展与新器官的生长,降低水分利用效率。复水通常可以减轻干旱对生长的限制,但科学家对叶片气孔限制与非气孔限制在光合能力恢复过程中的作用观点不同,这与植物种类、胁迫程度有很大关系(Galle et al.,2009;Hu et al.,2010)。Hu 等(2010)指出,中度干旱胁迫下气孔限制是多年生 C_3 草种光合能力恢复的限制因素。Varone(2012)比较了地中海地区不同种类植物幼苗在干旱—复水过程中的恢复能力,证明草本植物幼苗气孔限制是光合能力恢复的主要限制因素,而非气孔限制是树苗光合能力恢复的主要限制,草本植物幼苗的净光合速率较树苗恢复迅速。在未来气候变化趋势下,目前的研究对气孔限制与非气孔限制对旱后复水的恢复能力及作用还需要进一步了解。

大气 CO_2 浓度升高影响到生态系统中氮、磷、钾等元素的循环过程。氮素是作物生长发育的基本元素(Reich et al.,2006),当植物受到氮素限制时,叶片氮含量降低,氮利用效率下降。有研究认为,大气 CO_2 浓度升高可以增强植株氮素再利用性与可移动性(Dyckmans et al.,2002;Watanabe et al.,2011;Franzaring et al.,2011;Kim et al.,2011),尤其在干旱的生态系统中(Xu et al.,2007)。本研究从水分利用效率、光合潜力和生长潜力的角度进行分析,旨在解决干旱少雨与土壤贫瘠并存的生态系统中,玉米植株在干旱—复水过程中的适应与恢复能力对大气 CO_2 浓度升高的响应,对于阐明我国西北黄土高原干旱与半干旱地区 C_4 作物对未来气候变化的适应机制有重要意义。

8.2 试验设计与方法

挑选外形饱满的玉米种子(郑单 958)经 2% 的 NaOCl 溶液消毒 30 min,用蒸馏水反复冲洗,用 1 mM $CaCl_2$ 溶液浸种过夜,再置于 25℃恒温培养箱中萌发,每天补水,出苗大约 3 d。选取整齐均匀的发芽种子移至装有干净石英砂的塑料桶中,每桶种植 6 颗幼苗,塑料桶高 15 cm,直径上为 18 cm、下为 13 cm。并放入人工气候室(每个气候室控制面积 12 m^2,杭州求是人工环境有限公司,杭州)。将人工气候室 CO_2 浓度控制为 380 $\mu mol/mol$ 和 750 $\mu mol/mol$ 两个水平,每 3 d 更换一次 Hoagland 营养液。白天光照为 600 $\mu mol/(m^2 \cdot s)$,光暗周期为 12 h/12 h,昼夜温度为 25 ℃/20 ℃,昼夜空气相对湿度(RH)为 40%/55%;培养过程中剔除生长过快或过慢的植株,使待测植株长势基本一致。

第四部分 大气 CO_2 浓度升高对北方主要 C_4 作物的影响

从 80% 的幼苗完全展开第三片叶片开始,在营养液中加入 15% 的 PEG-6000（-0.4 M Pa）作为干旱处理,并将营养液按照氮素浓度分为低氮（5 mM）和正常（15 mM）两个级别。试验期间,每 3 d 更换 Hoagland 营养液,并将其 pH 值控制在 5.8～6.2。更换营养液时,用 1000 mL 蒸馏水将原有石英砂中营养液沿盆底小孔冲走,并在石英砂中加入新营养液。灌水于干旱胁迫的第 12 d 进行。气体交换参数（LI-6400 型光合作用系统,LI-COR,美国）与叶绿素荧光参数（PEA,Hansatech,英国）于复水后第 5 d 收集,相对生长速率于复水后第 6 d 测定。本研究中每次取样均设有 5 个重复,试验处理如表 8-1 所示。

表 8-1 试验处理名称与处理方式

CO_2 浓度（μmol/mol）	处理类型	氮素水平（mM）	水分状况	灌水强度（mL/桶）	处理名称
380	对照		水分充足	—	Control
	干旱胁迫（D）	15	15% PEG	—	D_0
				300	D_{300}
				600	D_{600}
				900	D_{900}
	氮素限制（N）	5	水分充足	—	N
	干旱胁迫+氮素限制（D-N）		15% PEG	—	D_0-N
				300	$D_{300}-N$
				600	$D_{600}-N$
				900	$D_{900}-N$

续表

CO_2浓度 (μmol/mol)	处理类型	氮素水平 (mM)	水分状况	灌水强度 (mL/桶)	处理名称
750	CO_2倍增	15	水分充足	—	C
	CO_2倍增 +干旱胁迫 (C+D)		15% PEG	—	C+D_0
				300	C+D_{300}
				600	C+D_{600}
				900	C+D_{900}
	CO_2倍增 +氮素限制 (C−N)	5	水分充足	—	C−N
				—	C+D_0−N
	CO_2倍增 +干旱胁迫 +氮素限制 (C+D−N)		15% PEG	300	C+D_{300}−N
				600	C+D_{600}−N
				900	C+D_{900}−N

8.3 结果与分析

8.3.1 复水对植株含水量的影响

图 8-1(a)表示了复水对不同处理植株含水量的影响,正常 CO_2 下植株复水后含水量高于 CO_2 倍增水平。复水对相对生长速率的刺激作用和对水分含量的刺激作用呈显著二次负相关关系(图 8-1(b)中:$PE_{RGR} = 0.25 + 123.41 PE_{WC} - 4579.75 PE_{WC}^2$;$R^2 = 0.7738, P < 0.01$),生长速率随着含水量提高而降低。

第四部分 大气 CO_2 浓度升高对北方主要 C_4 作物的影响

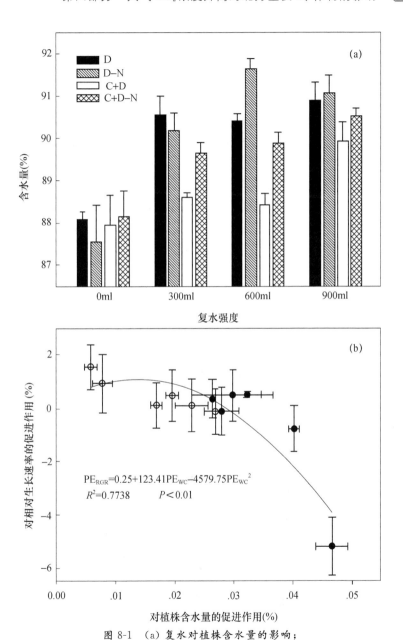

图 8-1 （a）复水对植株含水量的影响；
（b）复水对植株生长速率的增强与对植株含水量的增强之间的关系

8.3.2 植株相对生长速率的变化特征

PDL 表示前期干旱对植株生长的限制程度。图 8-2(a)从相对生长速率的角度表述了前期干旱对植株生长的限制。其中,C+D-N 处理显著提高了干旱对叶鞘相对生长速率的限制。D 和 D-N 处理显著提高了干旱对叶片相对生长速率的限制。D-N 处理下干旱对根系相对生长速率的限制最小。图 8-2(b)从生物量的角度表述了前期干旱对植株生长的限制。结果表明,D 处理显著增加了干旱胁迫对叶片与叶鞘生物量积累的限制,C+D 和 C+D-N 显著增加了干旱胁迫对根系生物量积累的限制。

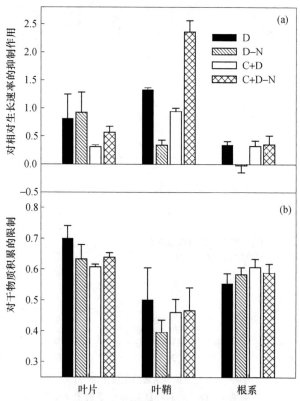

图 8-2　前期干旱(PDL,pre-drought limitation)对植株生长速率(a)和植株生物量(b)的限制

图 8-3 中 PE 表示复水后不同处理各器官相对增长速率较复水前的提高比率。复水没有提高正常 CO_2 浓度下受旱植株的相对生长速率,这与高 CO_2 浓度的结果相反,$C+D_{300}$ 和 $C+D_{600}$ 两个复水强度对植物相对生长速率提高最多。氮素限制下,$D_{300}-N$ 和 $D_{600}-N$ 两个水平的复水仅显著提高了叶片的相对生长速率,降低了根系与叶鞘的相对生长速率。CO_2 倍增条件下,$C+D_{300}-N$ 和 $C+D_{600}-N$ 处理提高了叶片与叶鞘的相对生长速率,对根系的生长没有提高效果。

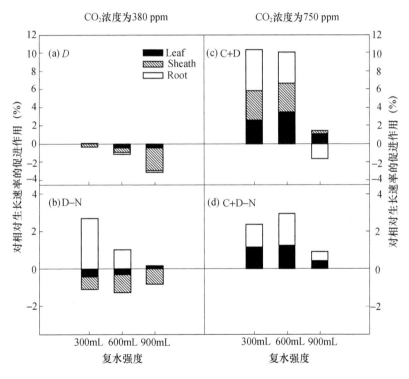

图 8-3 不同程度复水对植株生长速率的影响

(a) D 正常大气 CO_2 浓度下水分胁迫;(b) D-N 正常大气 CO_2 浓度下水分与氮素共同限制;
(c) C+D CO_2 倍增条件下水分胁迫;(d) C+D-N CO_2 倍增条件下水分与氮素共同限制。

8.3.3 光合潜力与植株含水量的关系

图 8-4 表示各处理复水后叶片光合性能与植株含水量之间的相关关系。其中,F_v/F_m、Φ_{PSII}、P_n 和 P_n/G_s 均与植株含水量呈显著二次正相关关系,最大值出现在 88.6% 与 89.87% 含水量之间。T_r 与植株含水量也呈显著正相关关系,最大值出现在 90.68% 含水量。NPQ 与植株含水量呈显著二次负相关关系。

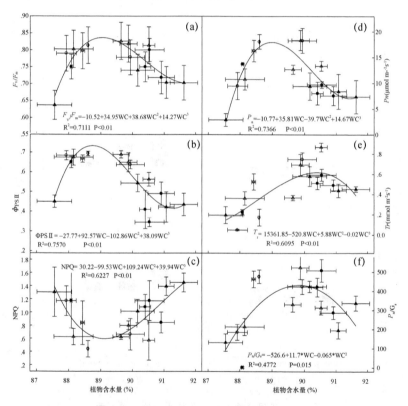

图 8-4 复水后叶片光合潜力与植株含水量之间的关系

(a) PSⅡ 最大光化学效率 (F_v/F_m);(b) PSⅡ 实际量子效率 (Φ_{PSII});(c) 非光化学淬灭系数 (NPQ);(d) 净光合效率 (P_n);(e) 气孔导度 (G_s);(f) 内在水分利用效率 (P_n/G_s)。○,C+D;△,C+D−N;●,D;▲,D−N。

第四部分 大气 CO_2 浓度升高对北方主要 C_4 作物的影响

如表 8-2 所示,复水显著改变了受旱植株的光合特性。复水后,CO_2 倍增较大气 CO_2 水平显著升高了植株净光合速率(P_n)、净光合速率/蒸腾速率(P_n/T_r)和净光合速率/气孔导度(P_n/G_s)值,其中 $C+D_{600}-N$ 使叶片 P_n 和 P_n/G_s 达到最高值,分别为 18.33 $\mu mol(CO_2)/(m^2 \cdot s)$ 和 522.699。$C+D_{300}$ 使 P_n/T_r 达到最大值 105.56。

表 8-2 前期干旱与复水过程中气体交换参数的变化

CO_2 浓度 ($\mu mol/mol$)	处理	P_n [$\mu mol/(m \cdot s)$]	P_n/T_r [($\mu mol/mmol$)]	P_n/G_s ($\mu mol/mol$)
	前期干旱			
380	D_0	0.01±0.12 c	0.02±4.62 d	0.26±0.02 c
	D_0-N	2.85±1.26 b	14.74±6.22 c	130.53±15.07 b
750	$C+D_0$	10.75±2.04 a	202.14±26.14 a	184.30±21.28 a
	$C+D_0-N$	9.44±2.63 a	29.87±7.04 b	212.86±24.58 a
	复水后			
380	D_{300}	9.70±0.64 d	13.01±5.96 e	507.32±58.59 a
	D_{600}	8.04±2.20 de	15.73±6.99 de	420.39±48.55 c
	D_{900}	7.56±1.82 e	15.47±6.95 de	287.23±33.17 e
	$D_{300}-N$	9.40±2.63 d	16.25±6.37 d	423.50±48.91 bc
	$D_{600}-N$	7.37±3.22 e	16.33±7.15 d	337.30±38.95 d
	$D_{900}-N$	8.40±3.37 d	19.21±8.05 d	194.25±22.43 f
750	$C+D_{300}$	18.05±1.44 a	105.56±18.67 a	475.31±54.89 ab
	$C+D_{600}$	16.27±2.09 ab	34.97±9.13 b	461.47±53.29 b
	$C+D_{900}$	18.27±2.43 a	30.81±9.05 b	422.59±48.80 bc
	$C+D_{300}-N$	12.67±0.88 c	34.64±7.63 b	330.03±38.11 d
	$C+D_{600}-N$	18.33±1.97 a	26.51±6.45 c	522.69±60.36 a
	$C+D_{900}-N$	13.33±0.95 c	26.03±7.80 c	313.54±36.21 d

8.3.4 复水对植株新叶生长的影响

如图 8-5(a)所示,正常氮素水平下,仅 CO_2 倍增条件下 $C+D_{300}$ 和 $C+D_{600}$ 处理极显著提高新叶生长速率。氮素限制条件下,两个 CO_2 水平中 $D_{300}-N$、$D_{600}-N$、$C+D_{300}-N$ 和 $C+D_{600}-N$ 均极显著提高新叶的相对生长速率。但是,最高复水量下 $D_{900}-N$ 和 $C+D_{900}-N$ 处理限制了新叶的生长。在正常 CO_2 水平下,复水对正常供氮植株(D 处理)的新叶生长速率没有显著影响。如图 8-5(b)所示,新叶相对生长速率与整株相对生长速率呈显著线性正相关关系($RGR_{new\ leaf}=38.38+0.54RGR_{plant}$; $R^2=0.4832$, $P<0.01$),因此,复水后新叶恢复生长的速度与整株生长的速度密切相关。

光系统Ⅱ(PSⅡ)的性能参数指示了新叶生长的潜在能力。正常 CO_2 水平下,正常供氮使植株新叶 F_m 随着复水强度升高而降低。$D-N$ 和 $C+D$ 处理使 F_m 在供水梯度 $0 \sim 600$ mL 升高,在供水梯度为 900 mL 时降低。CO_2 倍增水平下,不同水平的复水均提高了氮素限制植株的 F_m(图 8-6)。复水显著升高了正常 CO_2 浓度下氮素充足植株的 F_v/F_m,对其余处理没有影响。

8.4 讨论

在降雨量偏少且土壤贫瘠的干旱半干旱地区,了解植物生长对突发性降雨的应对策略在探索全球气候变化的影响中具有重要意义(Swemmer et al.,2007)。大多数情况下,干旱胁迫制约植物的生长能力,复水可以激活植株的生长潜力(Hu et al.,2010;Xu et al.,2009)。本研究表明,在正常与倍增的大气 CO_2 浓度下,复水均可以增加玉米植株的含水量。适当的植株含水量使其光合能力和生长速率的恢复加快,含水量过高(多于 90.5%)则制约植株生长。CO_2 倍增可提高复水后玉米幼苗的光合能力和新叶生长速率,使植株具有较高的生长潜力。

第四部分　大气 CO_2 浓度升高对北方主要 C_4 作物的影响

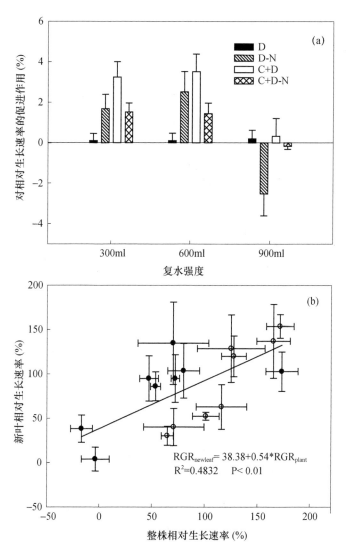

图 8-5　(a) 复水对新叶生长速率的增强效应；
(b) 新叶相对生长速率与整株相对生长速率之间的关系
○：CO_2 浓度为 750 ppm；●：CO_2 浓度为 380 ppm

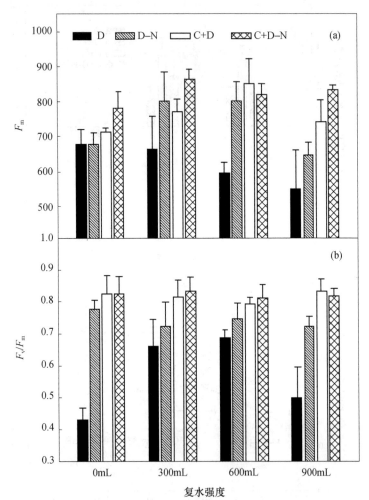

图 8-6 不同程度复水后玉米新叶的光合潜力变化
(a)叶绿素 a 最大荧光 (F_m);(b) PSⅡ 最大光化学效率 (F_v/F_m)

8.4.1 复水对植株生长的影响

在正常 CO_2 浓度下,复水使玉米幼苗的植株含水量提高到

90.5%～91.0%。其中,复水 900 mL 使植株含水量达到最高,但抑制了叶鞘生长,降低了整株生物量积累,植株生长的适应能力显著下降(图 8-2(a),(b))。Robredo 等在大麦的研究中认为,高浓度 CO_2 可以加快大麦生长复水后的恢复速度(Robredo et al.,2011)。本研究中,CO_2 倍增条件下,3 个复水强度使植株的含水量维持在 88.4%～89.9%。其中,复水 300 mL 和 600 mL 可提高植株生物量积累,复水 900 mL 使植株含水量过多 (89.9%),限制根系生长,导致生物量降低。Wang 等(2010)指出,高浓度 CO_2 可以使受旱植株更大程度地增加根/冠比。本研究进一步认为,CO_2 倍增增强了植株根系在复水后的恢复能力,导致植株生长加快。

目前,高浓度 CO_2 可以减轻干旱胁迫对植株伤害的观点已被广泛认同(Reich et al.,2006;Albert et al.,2011)。但是,在氮素缺乏地区,复水对干旱胁迫的缓解程度在高浓度 CO_2 下的响应还鲜有报道(Markelz et al.,2011)。本研究中,两个 CO_2 浓度下,复水均增加了缺氮植株的根系生产力。即便如此,在正常 CO_2 浓度下,复水并没有增强缺氮植株的叶鞘生长于叶面积扩展。CO_2 倍增则反之,复水后缺氮植株叶面积迅速扩展,这可能是由植株内在的高水分利用效率引起的(图 8-4(f))。

前人研究证明,高浓度 CO_2 可以增强植株生长中氮的有效性与可移动性(Franzaring et al.,2011;Kim et al.,2011),提高氮素代谢能力,促进复水后生长能力的恢复(Robredo et al.,2011)。本研究也证明 CO_2 倍增增强了氮素有效性(第 5 章),显著促进了复水后缺氮植株的叶片生长。

8.4.2 复水对植株光合能力的影响

在复水后光合能力恢复的过程中,光化学(例如 Φ_{PSII} 和 F_v/F_m)与气孔的恢复能力因植物种类和水分状况而不同(Galle et al.,2007;Hu et al.,2010)。本研究与前人结果类似,F_v/F_m 在干旱与复水过程中的变化趋势与 P_n 相同(图 8-4),光合能力的恢复显著依赖于光化学能力(Hu et al.,2010)。此外,CO_2倍增较正常CO_2条件使植株在复水后具有较高的F_v/F_m和Φ_{PSII},较低的NPQ,证明CO_2倍增可以增强叶片对光能的捕获与转化能力,提高量子效率。

本研究中复水后光合能力恢复与植株含水量显著相关。在大气CO_2水平下,复水后随着含水量的升高,P_n下降幅度大,T_r和G_s下降幅度较小,引起P_n/T_r和P_n/G_s降低(图8-4,表8-2),导致含水量较高的植株光合能力较弱。然而,在CO_2倍增条件下,随着复水强度的升高,植株含水量提升较小,而P_n的恢复水平高于T_r和G_s,使叶片具有较高的P_n/T_r和P_n/G_s,水分利用效率提高,用较少的水分固定较多的CO_2。这与Ephrath在棉花上的研究结论相同(Ephrath et al.,2011)。

复水过程中,气孔的开闭能力往往限制了光合能力的恢复(Hu et al.,2010)。Markelz(2011)认为,氮素胁迫可以增强干旱条件下的气孔限制,高浓度CO_2有助于延缓并减轻这种限制。本研究进一步证明,CO_2倍增水平下,受氮素与水分共同胁迫的植株较正常CO_2浓度显著提高了叶片T_r和G_s,并减轻气孔限制,增强P_n。复水600mL和900mL提高了CO_2倍增水平下缺氮植株的P_n/T_r和P_n/G_s,增强了水分利用效率,使植株在较小的气孔导度下具有较高的光合能力,降低了气孔开闭在光合恢复过程中的作用。

8.4.3 复水对新叶生长的影响

植株的生长速率与新器官的生长有密切联系(Nure et al.,2010;Xu et al.,2009;Pinheiro et al.,2004)。在复水过程中,植株新叶生长越快整株相对生长速率越高(图8-5(b))。CO_2倍增使复水植株新叶生长能力增强,光合能力较快得到恢复(F_m和F_v/F_m)。

然而,氮素限制条件下,两种CO_2水平均增强了中低等复水强度(300 mL和600 mL)下植株的新叶生长。虽然,复水强度过大(900 mL)限制了新叶生长,但高强度复水对新叶生长的限制程度随

CO_2 浓度升高而降低(图 8-5(a)),所以 CO_2 浓度升高有助于减轻复水过量对新叶生长的限制(图 8-2(b)与(d))。因为氮素循环与蛋白质的合成在新器官生长中具有重要作用(Lattanzi,2005),高浓度 CO_2 使氮素的可利用性与可移动性增强(Franzaring et al.,2011; Kim et al.,2011),所以新器官生长能力提高,即使在氮素缺乏的环境中也是如此。

CO_2 倍增水平对复水后新叶 F_v/F_m 恢复没有显著优势(图 8-6),但显著提高了 F_m,证明复水后短期光能吸收能力优于光能转化能力先恢复。正常 CO_2 水平下,氮素缺乏的植株在适量复水(300 mL 和 600 mL)后新叶 F_m 的恢复较快,氮素充足的植株则反之,说明氮素缺乏可加快植株光能吸收能力恢复。

8.5 小　　结

本研究中,复水后植株相对生长速率与含水量呈显著负相关,含水量适量增加促进光合生理与生长恢复,含水量过高则不利于生长。正常 CO_2 水平下,复水后植株含水量较高,光合能力较弱,新叶生长潜力低,生长受到抑制。CO_2 倍增水平下,复水使植株含水量适量增长,叶片光化学能力增强、水分利用效率提高,新叶生长快,导致生物量积累迅速。所以,在高浓度 CO_2 条件下,玉米植株对干旱—复水过程具有较高的适应能力,即使在氮素缺乏条件下也是如此。

第 9 章　大气 CO_2 浓度升高对糜子光合特性与产量的影响

9.1　引言

工业革命以来，全球大气 CO_2 浓度从 280 μmol/mol 上升至 406 μmol/mol，预测在 21 世纪末将上升至 421~936 μmol/mol(IPCC，2013)。大气 CO_2 浓度升高引起大气温度升高、全球降水格局改变，进而导致作物产量降低(Thomson et al.，2005)。但也有研究认为，大气 CO_2 浓度升高可以促进作物生产力提高，抵消气候变化引发的负面影响(Leakey et al.，2006)。

与 C_3 植物相比，C_4 植物叶片具有较高的光合碳同化效率(Liu et al.，2013)。美国 BioCON 平台(biodiversity，CO_2，nitrogen，位于美国明尼苏达州的贝瑟尔大学)的研究表明，C_4 野草在饱和光强下的净光合速率在大气 CO_2 浓度持续升高的环境中没有显著增强，但是一项对在美国伊利诺伊州和亚利桑那州马里科帕 FACE 平台 15 年研究的 meta 分析表明，高粱等 C_4 作物的生物量积累在大气 CO_2 浓度升高下显著增长了 15%~20%(2005)。有研究认为，C_4 植物对大气 CO_2 浓度升高条件相应的差异与物种间 CO_2 的饱和点不同有关(Wand et al.，1999)。

Wand 等(1999)报道，C_3 与 C_4 草本植物的 G_s 在大气 CO_2 浓度升高环境中平均降低了 20%，且 C_3 与 C_4 之间没有差异。C_3 与 C_4 植物的潜在水分利用效率(WUE_i)(P_n/G_s)存在较大差异。在 FACE 条件下，C_3 植物的 WUE_i 增长了 68%。在美国马里科帕 FACE 试验站的研究表明，大气 CO_2 浓度升高可以改善高粱水分状况(Conley et al.，2001；Wall et al.，2001)，提升 WUE。伊利诺伊 FACE 试验站的研究表明，大气 CO_2 浓度升高下玉米叶片的 G_s 降低了 34%，进而减少了耗水（Leakey et al.，2006）。也有研究表明，谷子和高粱的 G_s 与 T_r 降低，WUE 提高（Vu et al.，2009；郝兴宇等，2010）。所以，大气 CO_2 浓度升高可能会改变大多数 C_4 植物的 WUE，增强抗旱性。

在 FACE 平台中，前人比较了不少 C_3 植物与 C_4 植物对大气 CO_2 浓度升高的响应特征（Ainsworth et al.，2005），目前对 C_4 作物适应机制的研究较少。糜子（*Panicum miliaceum* L.）是世界上最古老的作物之一，中国最早在 7000 年前开始种植（Underhill，1997）。由于 C_4 植物发育过程中需要消耗较多的水分和营养物质，所以较高的水分和养分利用效率有利于使糜子适应干旱与贫瘠的土壤条件。糜子生育期较短，对水分和养分需求较低，所以在中国、中亚与俄罗斯等地的干旱半干旱与贫瘠土壤中广泛种植。目前，中国糜子的年种植面积将近 1000 亩*，主要种植于北方山区（Dai et al.，2011；Hunt et al.，2011）。水稻（Seneweera et al.，2002）、玉米（Leakey et al.，2006）、大豆（Hao et al.，2012；Bernacchi et al.，2006）和小麦（Khalil et al.，2013）等大宗作物由于较高的经济价值，在未来气候变化中受到较多的关注。糜子栽培历史悠久，抗逆性强（Hunt et al.，2011），但在未来气候条件下的适应性研究较少。大气 CO_2 浓度升高环境下，糜子植株在光合特性、生长和产量形成的直接与间接变化均会对低收入山区的经济收入带来较大影响。同时，作为 C_4 植物，糜子对大气 CO_2 浓度持续升高环境的响应和适应机制可能与 C_3 大宗作物差异较大。

本研究主要关注：(1)在大气 CO_2 浓度升高条件下，糜子的生理、叶绿素荧光与产量变化之间的关系；(2)大气 CO_2 浓度升高是否会降低糜子的 G_s，进而改变 WUE？本研究有助于理解中国北方 C_4 作物对未来气候变化的适应性。

---

* 1 亩=1/15 hm^2，下同。

9.2 试验设计与方法

本试验于 2013—2014 年在山西农业大学人工模拟控制 CO_2 浓度的开顶式气室(OTC)进行。生长室长 6 m,宽 4 m,高 3 m,顶开口呈八角形。OTC 生长室内地面中部安装带风扇的 CO_2 发散器,地面埋有与控制室 CO_2 钢瓶相连的通 CO_2 气体的管道,各生长室安装 CO_2 浓度传感器及水分测定器,可随时监测室内 CO_2 浓度,高 CO_2 浓度室内 CO_2 气体由减压阀控制的压缩 CO_2 钢瓶供给。处理期间的 CO_2 浓度由钢瓶流出,进出温室中由通气小管散开,风扇运行,保证室内 CO_2 浓度均匀,设置高 CO_2 浓度处理的目标浓度,检测仪连续监测,当室内 CO_2 浓度低于目标浓度时,自动将 CO_2 注入室内,使其浓度保持在 (560 ± 20) μmol/mol,通气时间 8:00—18:00,10 min 记录一次数据,正常 CO_2 浓度室内只通空气,不通入 CO_2。

糜子于 6 月 20 日播种于塑料种植箱中,箱长 60 cm,宽 40 cm,高 35 cm,在箱内部依次铺沙土、壤土共 28 cm 深。每个箱子播种 15 穴,每穴 5 粒种子,出苗后留 1 株健壮的幼苗。于糜子生长的关键生育期测定气体交换参数与光合特征曲线(LI-6400 型光合作用系统,LI-COR,美国)、叶绿素荧光特性(PAM-2500 便携式荧光调制解调器,WALZ,德国),于成熟期测定谷子产量构成。

9.3 结果与分析

9.3.1 气体交换参数

大气 CO_2 浓度升高条件下,糜子旗叶的 P_n 在 2013 年的抽穗初期与灌浆期分别提升了 33.2% 和 80.3%,在 2014 年分别提升了 46.7% 和 80.3%。G_s 拔节期与灌浆期在 2013 年分别提升了 11.1% 和 24.2%,在 2014 年提升了 25.0% 和 50.0%。C_i 在 2013 年和 2014 年的拔节期与灌浆期分别提高了 96.8%、13.1% 与 93.8%、91.1%。大气 CO_2 浓度升高下,T_r 在 2013

第四部分 大气 CO_2 浓度升高对北方主要 C_4 作物的影响

年和 2014 年分别平均提升了 12.0% 和 21.2%，WUE_i 分别提升了 42.2% 和 18.5%（表 9-1）。

表 9-1 大气 CO_2 浓度升高对糜子第一片完全展开叶气体交换特征的影响

年份	生育期	Growth [CO_2]	P_n [μmol/ (m²·s)]	G_s [mol(H_2O)/ (m²·s)]	C_i [μmol (CO_2)/ mol]	T_r [mmol (H_2O)/ (m²·s)]	WUE_i [μmol (CO_2)/mmol (H_2O)]
2013	抽穗初期	CK	13.66± 0.54	0.09± 0.01	112.21± 3.50	1.74± 0.07	159.71± 2.45
		E[CO_2]	18.19± 0.71	0.11± 0.01	220.80± 8.52	1.91± 0.10	178.69± 5.65
	灌浆期	CK	10.04± 0.69	0.07± 0.00	168.00± 4.52	2.01± 0.05	113.15± 13.29
		E[CO_2]	18.10± 0.30	0.09± 0.00	190.00± 20.87	2.29± 0.03	209.21± 2.79
		p_{CO_2}	0.00	0.01	0.00	0.02	0.00
2014	抽穗初期	CK	19.01± 0.99	0.12± 0.01	110.57± 4.51	2.26± 0.13	154.01± 2.31
		E[CO_2]	27.89± 1.08	0.15± 0.01	214.30± 11.87	2.44± 0.15	194.85± 7.09
	灌浆期	CK	12.78± 0.26	0.08± 0.00	114.75± 4.06	1.75± 0.03	159.02± 2.73
		E[CO_2]	20.93± 0.61	0.12± 0.00	219.25± 6.30	2.42± 0.06	176.02± 4.10
		p_{CO_2}	0.00	0.03	0.00	0.04	0.04

注：E[CO_2]表示 CO_2 浓度升高的处理；CK 为对照处理，全书后同。

9.3.2 叶绿素荧光特性

大气 CO_2 浓度升高下，F_v/F_m 在 2013 年降低了 3.2%，但在 2014 年没有显著变化。F_v'/F_m' 在 2013 年降低了 9.3%，但在 2014

年提升了 5.2%。Φ_{PSII} 在 2013 年和 2014 年分别平均提高了 51.3% 和 28.3%,qP 分别平均提高了 67.5% 和 12.3%,NPQ 分别平均降低了 11.3% 和 19.5%(表 9-2)。

表 9-2　大气 CO_2 浓度升高对糜子第一片完全展开叶叶绿素荧光特征的影响

年份	生育期	Growth [CO_2]	F_v/F_m	F_v'/F_m'	Φ_{PSII}	qP	NPQ
2013	抽穗初期	CK	0.80±0.01	0.49±0.02	0.20±0.01	0.42±0.04	1.53±0.04
		E[CO_2]	0.78±0.00	0.50±0.01	0.31±0.02	0.63±0.03	1.33±0.08
	灌浆期	CK	0.76±0.01	0.48±0.02	0.19±0.01	0.41±0.02	1.30±0.08
		E[CO_2]	0.73±0.01	0.38±0.03	0.28±0.02	0.76±0.07	1.18±0.07
		p_{CO_2}	0.00	0.03	0.00	0.00	0.01
2014	抽穗初期	CK	0.79±0.00	0.41±0.03	0.29±0.04	0.72±0.08	2.58±0.23
		E[CO_2]	0.80±0.00	0.49±0.01	0.41±0.02	0.83±0.04	1.78±0.17
	灌浆期	CK	0.77±0.01	0.38±0.03	0.23±0.03	0.59±0.04	1.86±0.29
		E[CO_2]	0.78±0.00	0.40±0.01	0.26±0.01	0.64±0.02	1.79±0.15
		p_{CO_2}	0.08	0.02	0.01	0.07	0.01

注:表中 F_v/F_m 为 PSⅡ 最大光化学量子产量;F_v'/F_m' 为 PSⅡ 有效光化学量子产量;Φ_{PSII} 为 PSⅡ 实际光化学效率;qP 为光化学淬灭系数;NPQ 为非光化学淬灭系数。

9.3.3　光合特征曲线

对 P_n/C_i 响应曲线拟合的结果表明(图 9-1),P_n 随 C_i 提升而升高,在两种 CO_2 环境中没有差异。g_s 随 C_i 提升而降低。但在大气 CO_2 浓度升高条件下,当糜子叶片的 C_i 在 200~250 μmol/mol 范围内时,G_s 降低幅度较小。

第四部分 大气 CO_2 浓度升高对北方主要 C_4 作物的影响

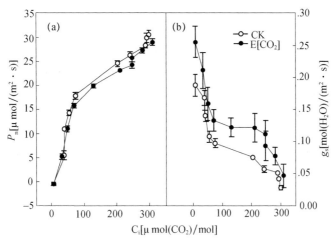

图 9-1 大气 CO_2 浓度升高对糜子旗叶 P_n/C_i 与 G_s/C_i 特征曲线的影响

9.3.4 形态特征

大气 CO_2 浓度升高条件下,糜子植株株高在 2013 年和 2014 年分别升高了 5.4% 和 5.7%。节间数在 2014 年升高 11.1%,2013 年没有显著变化。大气 CO_2 浓度升高下糜子穗长没有显著变化。茎直径在 2013 年和 2014 年分别升高了 20.4% 和 26.7%(表 9-3)。

表 9-3 大气 CO_2 浓度升高对糜子形态指标的影响

年份	处理	株高(cm)	节数	穗长(cm)	茎粗(cm)
2013	CK	112.66±2.63	7.03±0.11	25.36±0.68	0.54±0.01
	E[CO_2]	118.79±1.99**	6.97±0.18	27.97±0.76	0.65±0.01**
2014	CK	136.58±3.06	6.92±0.14	26.42±0.97	0.45±0.02
	E[CO_2]	144.35±2.22*	7.69±0.10**	26.69±0.76	0.57±0.02**

9.3.5 生物量和产量

大气 CO_2 浓度升高下,叶、茎、穗和地上部生物量在 2013 年分别显著增加了 29.9%、41.6%、36.3%和 37.3%,在 2014 年分别显著增加了 56.0%、35.9%、12.9%和 23.8%。千粒重没有明显变化。每株粒数在 2013 年和 2014 年分别显著增加了 31.8%和 12%。每平方米产量在 2013 年和 2014 年分别提高了 34.2%和 13.6%(表 9-4,图 9-2)。

表 9-4 大气 CO_2 浓度升高对糜子器官干重和产量构成因素的影响

年份	处理	叶干重 (g/m²)	茎干重 (g/m²)	穗干重 (g/m²)	千粒重 (g)	每株粒数 (个)
2013	CK	72.5±4.58	220.4±21.67	420.8±9.17	7.1±0.08	1219.8±49.57
2013	E[CO_2]	94.2±1.25**	312.1±22.08*	573.7±35.83**	7.3±0.13	1607.9±87.37*
2014	CK	48.3±2.92	176.3±24.17	332.9±9.17	7.7±0.48	801.6±39.37
2014	E[CO_2]	75.4±6.25**	239.6±13.33*	375.8±10.83*	7.7±0.45	898.0±51.93*

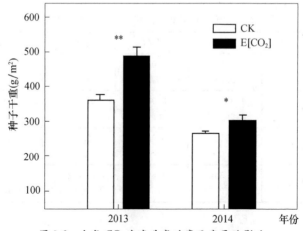

图 9-2 大气 CO_2 浓度升高对糜子产量的影响

9.4 讨论

有前人研究认为,大气 CO_2 浓度升高环境对 C_3 植物光合性能的提升幅度大于 C_4 植物(Jin et al.,2019;Xie et al.,2015)。本研究中,糜子的 P_n 在 CO_2 浓度升高环境下显著升高。不同类型的 C_4 植物光合特性在大气 CO_2 升高下的适应机制不同(Leakey et al.,2009)。比如,Flaveria trinervia 未完全展开叶的维管束鞘对 CO_2 的固定量占总叶片比例的 10%,但并没有参与 CO_2 浓缩机制(Moore Brandon et al.,1986)。大气 CO_2 升高增加了维管束鞘在光合碳通过过程中的作用,因而高粱叶片 P_n/C_i 曲线的初始斜率与 CO_2 饱和点都降低了(Jenny et al.,2000)。这与糜子完全展开叶的光合作用适应过程不完全一致。本研究中,糜子的初始斜率在 CO_2 浓度升高环境中并没有发生变化。前人研究表明,一些 C_4 植物的幼叶在高 CO_2 条件下光合性能会像 C_3 作物一样增强(Zong et al.,2014;Cousins et al.,2001)。在 CO_2 升高条件下,糜子旗叶的 P_n 在拔节期和灌浆期均高于正常 CO_2 条件。在正常 CO_2 浓度下,C_i 低于 P_n/C_i 曲线的饱和点,因而 CO_2 浓度升高时,P_n 增加(Watling et al.,1997;Ziska et al.,1997)。在 CO_2 浓度升高环境下,C_4 植物的光合性能会增加,因为在正常 CO_2 条件下光合速率并未达到饱和(Wand et al.,1999)。本研究中 P_n/C_i 曲线的结果表明,P_n 随着 C_i 增加而升高,但糜子的 P_n 没有达到饱和点,大气 CO_2 浓度升高会引起 P_n 上调。G_s/C_i 曲线表明,大气 CO_2 浓度升高下 G_s 高于对照,特别是 C_i 处于 200~250 μmol/mol 时,说明较高的 C_i 也许是糜子光合能力增强的原因之一。

大气 CO_2 浓度升高条件降低了 2013 年糜子叶片的 F_v/F_m 和 F_v'/F_m',但 2014 年没有显著变化,这可能是引起光合性能在 2013 年改变的原因之一。两年之间的差异可能与气候因素有一定关系,如均温、相对湿度、太阳辐照度等。大气 CO_2 浓度升高提高了糜子叶片的 Φ_{PSII} 和 qP,但降低了 NPQ。NPQ 的降低表示糜子叶片在电子传递较热耗散分配了更多能量。本研究中，Φ_{PSII}、qP、NPQ 与 P_n 的变化规律相一致。但 Leakey 等（2006）对玉米的研究表明，P_n、Φ_{PSII}、F_v'/F_m'、qP 与 NPQ 在大气 CO_2 浓度升高条件下没有显著变化。而 C_3 植物绿豆的 F_v'/F_m'、Φ_{PSII} 与 qP 显著降低（Gao et al.，2015），小麦（Tausz-Posch et al.，2013）与菘蓝（Hao et al.，2013）的 P_n、Φ_{PSII} 与 qP 升高，NPQ 降低，本研究结果与 C_3 植物的结果类似。所以无论是 C_3 还是 C_4 植物，不同种类的植物对于 CO_2 浓度升高的响应差异较大。

前人研究表明，大气 CO_2 浓度升高环境中，K^+ 通道的外流活性增加而内流活性降低，进而导致 s 型阴离子通道活性增强，激发保卫细胞中 Cl^- 释放，Ca^{2+} 浓度增加（Hanstein et al.，2002；Raschke et al.，2003；Ainsworth et al.，2007）。这些变化使大气 CO_2 浓度升高下保卫细胞的膜电位去极化，导致气孔关闭（Ainsworth et al.，2007）。较多研究证明，C_3 与 C_4 草本植物的 G_s 在 CO_2 浓度升高下均降低（Ainsworth et al.，2005；Leakey et al.，2010；Ainsworth et al.，2007；Gao et al.，2015）。相反，树木的 G_s 没有显著变化（Saxe et al.，1998）。火炬松的保卫细胞对大气 CO_2 浓度升高不敏感（Ellsworth，1999）。然而，本研究表明糜子 G_s 在大气 CO_2 浓度升高条件下升高。G_s/C_i 曲线结果显示，在 CO_2 浓度升高环境中 G_s 随 C_i 增加的下降幅度小于在正常 CO_2 环境中，尤其当 C_i 处于 $200\sim250$ $\mu mol/mol$ 时。也有研究表明，在 CO_2 浓度升高条件下，大豆（Bernacchi et al.，2006）、绿豆（Gao et al.，2015）、玉米（Leakey et al.，2006）的 G_s 降低。气孔适应可能是大气 CO_2 浓度升高下气孔导度升高的原因之一。

虽然在大气 CO_2 浓度升高下糜子叶片的 G_s 增加，光合能力改变，但 T_r 与 G_s 同时得到提高，说明整株植株的水分平衡在大气 CO_2 浓度升高环境下没有发生改变，这与前人的研究结果不一致（Ainsworth et al.，2005；Ainsworth et al.，2007；Hao et al.，2013）。虽然由于 P_n 的提高，糜子 WUE_i 也得到提升，但是较高的 T_r 意味着未来气候条件下糜子可能会面临水分亏缺问题。

第四部分　大气 CO_2 浓度升高对北方主要 C_4 作物的影响

9.5　结论

大气 CO_2 浓度升高条件下,糜子旗叶 P_n、G_s、C_i、T_r、WUE_i、qP 和 Φ_{PSII} 均得到提升。地上部生物量(叶、茎和穗)与产量同时也得到提高。每株穗数的增加使产量也得到增加。所以,糜子通过增加 G_s 而提高了大气 CO_2 浓度升高条件下的光合性能,进而导致产量增加。

第 10 章　大气 CO_2 浓度升高对谷子光合特性与产量的影响

10.1　引言

　　大气 CO_2 浓度升高将直接影响作物生长发育与产量形成（Zong et al.，2016；Soba et al.，2020）。但是，作物产量形成对大气 CO_2 浓度升高的响应特征因作物种类而异（Jin et al.，2019）。例如，C_3 作物小麦、水稻和大麦的产量将升高 19% 左右，C_3 豆科作物大豆、桃树、花生和芸豆产量升高约 16%，但 C_4 作物高粱则表现出下降趋势（Kimball et al.，2016）。C_4 作物是全球农业产量构成中的重要组成部分（Steffen et al.，2003）。谷子在中国具有悠久的种植历史，现在仍然是中国重要的作物之一，且具有较高的营养价值与抗旱性、抗瘠薄性（Lata et al.，2013；Diao et al.，2014；Goron et al.，2015）。同时，谷子由于较小的基因组（490 Mbp）、较小的个体体积和较快的世代繁殖时间，成为经典的 C_4 模式作物（Diao et al.，2014；Bennetzen et al.，2012；Jia et al.，2013）。然而，目前对大气 CO_2 浓度升高下谷子的响应特征研究较少。

　　前人对大气 CO_2 浓度升高下 C_3 植物的响应机制研究较多（Ainsworth et al.，2005；Seneweera et al.，2005），主要表现为光合能力增强、气孔导度降低，以及光呼吸减少（Aranjuelo et al.，2011；Thomson et al.，2005；Ainsworth et al.，2010）。前人研究表明，在大气 CO_2 浓度升高条件下，C_4 植物的生物量积累也表现为上调，但这并不是 CO_2 对光合作用的直接作用所引起的（Chun et al.，2011），因为 C_4 植物的光合性能在正常大气 CO_2 浓度下已经接近饱和（Ghannoum et al.，2000a）。C_4 作物的产量可以在光合性能没有增加的情况下表现出明显的增加趋势（Leakey et al.，2004；Leakey et al.,2006;Leakey and A.,2009)。有研究表明,大气 CO_2 浓度升高仅仅可以促进干旱胁迫条件下 C_4 植物的生长(Leakey et al.,2006;Markelz et al.,2011),对正常灌水条件无显著影响,但也有研究认为对两种水分条件下的 C_4 植物生长均能产生上调作用(Zong et al.,2016)。因此,前人对 CO_2 浓度升高下 C_4 植物响应机制的认识仍存在争议。

目前,前人对 CO_2 升高条件下植物分子适应机制的研究主要集中在杨树(Gupta et al.,2010;Taylor et al.,2005)、拟南芥(Li et al.,2006;Li et al.,2008a)、大豆(Ainsworth et al.,2006)、甘蔗和水稻(Gesch et al.,2003)上。在 CO_2 升高条件下,较多植物的光合能力表现为下调趋势(Seneweera et al.,2005;Li et al.,2008a;Salvucci et al.,2004;Kontunen-Soppela et al.,2010),且光捕获基因下调(Li et al.,2008a;Cseke et al.,2009;Takatani et al.,2014)。

大气 CO_2 升高下,植物叶片的光合能力下调与 RbcS 蛋白(Rubisco 小亚基)的转录过程被抑制有关(Gesch et al.,2003;Leakey et al.,2009;Takatani et al.,2014)。在大气 CO_2 浓度升高条件下,许多叶绿体有关的功能基因的转录丰度降低(Li et al.,2006),但是与光捕获、发育、防御以及信号传导等功能有关基因的转录却上调 Gupta2005(Gupta et al.,2010;Li et al.,2008a)。几种编码叶绿体转运蛋白与糖类转运蛋白的基因转录水平上调,并且编码线粒体转运蛋白的基因转录丰度也增加(Leakey et al.,2009;Souza et al.,2010)。Shi 等(2015)研究表明,依次合成的依赖 NADPH 氧化酶的 H_2O_2 与依赖 NR 的 NO 可以在 OST1 下游发挥作用,并参与了 CO_2 浓度升高引发的气孔关闭过程。CO_2 浓度升高下,为维持糖酵解、线粒体电子传递与三羧酸循环,呼吸系统相关基因将表现出上调趋势(Fukayama et al.,2011)。本研究为探讨 CO_2 浓度升高下谷子的生理和分子适应机制做出如下假设:(1)叶片光合生理、叶绿素荧光特征与产量是否会发生变化。(2)光合相关基因表达是否会发生改变。(3)基因表达变化与光合生理、产量之间的关系是什么？本研究将有助于进一步理解 C_4 谷物在 CO_2 浓度升高下的调节适应机制。

10.2 试验设计与方法

本试验于 2014—2015 年在山西农业大学人工模拟控制 CO_2 浓度的开顶式生长室(OTC)进行。生长室长 6 m,宽 4 m,高 3 m,顶开口呈八角形。OTC 生长室内地面中部安装带小管辐射状的 CO_2 发散器,地面埋有与控制室 CO_2 钢瓶相连的通 CO_2 气体的管道,各生长室安装 CO_2 浓度传感器及水分测定器,可随时监测室内 CO_2 浓度,高 CO_2 浓度室内 CO_2 气体由减压阀控制的压缩 CO_2 钢瓶供给。处理期间的 CO_2 浓度由钢瓶流出,进出温室中由通气小管散开,风扇运行,保证室内 CO_2 浓度均匀,设置高 CO_2 浓度处理的目标浓度,检测仪连续监测,当室内 CO_2 浓度低于目标浓度时,自动将 CO_2 注入室内,使其浓度保持在 (560 ± 20) $\mu mol/mol$,通气时间 08:00—18:00,10 min 记录一次数据,正常 CO_2 浓度室内只通空气,不通入 CO_2。

谷子于 5 月 8 日播种于塑料种植箱中,箱长 60 cm,宽 40 cm,高 35 cm,在箱内部依次铺沙土、壤土共 28 cm 深。每个箱子播种 15 穴,每穴 5 粒种子,出苗后留 1 株健壮的幼苗。于谷子生长的关键生育期测定气体交换参数(LI-6400 型光合作用系统,LI-COR,美国)、叶绿素荧光特性(PAM-2500 便携式荧光调制解调器,WALZ,德国),于成熟期测定谷子产量构成。

2014 年取谷子开花期的旗叶进行高通量测序分析。剪取叶片后迅速用液氮冷冻,再置于 $-80°C$ 冰箱保存。样品制备和转录组测序在深圳华大基因公司完成,基因表达水平通过 RPKM 方法计算(Mortazavi et al.,2008),用(Audic et al.,1997)的方法鉴定表达差异的基因,用 GO 和 KEGG 方法进行通路丰度分析。设计特异性引物(表 10-1),并用 RT-PCR 技术验证表达差异基因的实际表达量。

第四部分 大气 CO_2 浓度升高对北方主要 C_4 作物的影响

表 10-1 用于实时定量 PCR 基因表达研究的基因特异性引物序列

基因登录号	基因名称	通路	引物序列
Si014344	Caffeoyl-CoA O-methyltransferase	Phenylpropanoid biosynthesis	F ACGTGGGGGCGTTCGAC R TGAGGTCCCTGATGGCGG
Si030198	Protein phosphatase Type I inositol	Plant hormone signal transduction	F TCCTCGGACCACAAGCCC R GCCCTCCCAGAAGATGACG
Si000717	polyphosphate 5-phosphatase, arath	Inositol phosphate metabolism	F TCGTGAGCAAGCAGATGGT R GGTGGCAGCACACAAAGC
Si035574	Glycosyl hydrolases family	Plant hormone signal transduction	F GCCTACAACGACTACTACCA R CGAAAACGCAAGACCCTGA
Si036462	Peroxidase	Phenylpropanoid biosynthesis	F CTCACACGTTTGGCAGGGTA R GCGATAGGAATGCTCGGTAA
Si000774	NADP+-dependent malic enzyme	Carbon fixation in photosynthetic organisms	F TTGCTCAGCAGGTCTCAGAA R CAGCGGTAGTTGCGGTAAA
Si013194	Lipoxygenase	Linoleic acid metabolism	F AAGGAGATTGAGGGGATCAT R GTCACGCCTTCCTGAGAGA
Si034948	Phosphoglycerate mutase	Glycolysis/Gluconeogenesis	F TGAGCAAGTGGGTGGCATT R TCCTTGTCACGGAGCGGT
Si035904	α-galactosidase/ alpha-n-acetylgalactosaminidase	Galactose metabolism; Glycosphingolipid biosynthesis; Sphingolipid metabolism; glycerolipid metabolism	F CTGCACAAGACGCTGGACA R CCTGGACTTGAGCACGAACAT
Si010446	Leucine-rich repeat-containing protein	Plant-pathogen interaction	F GGAACCCGCTGGTGTGTC R GCCGTGAAGGTGCCGTA

10.3 结果与分析

10.3.1 产量构成与地上部生物量

大气 CO_2 浓度升高对两年的谷子植株生长和发育时期长短均没有显著影响,但显著改变了产量和地上部生物量积累。大气 CO_2 浓度升高条件下,2014 年和 2015 年谷子产量分别提高了 32% 和 11%,地上部生物量分别显著提高了 19% 和 8%。同时,相较于正常大气 CO_2 浓度,穗和叶干重在 2014 年分别提高了 18% 和 19%,在 2015 年分别提高了 20% 和 4%($P<0.05$)。同时,每株粒数 2014 年提高了 25%,2015 年提高了 8%,千粒重也显著提高。大气 CO_2 浓度升高下,茎直径在 2014 年增加了 16%,2015 年增加了 12%($P<0.01$)。分蘖数在 2014 年提升了 27%,2015 年提升了 14%($P=0.01$)。株高在 2014 年和 2015 年均得到提高,但穗长的变化在两年间表现不一致(表 10-2、表 10-3、图 10-1、图 10-2)。

表 10-2 大气 CO_2 浓度升高对谷子各器官生物量的影响

年份	处理	穗重 (g/m²)	叶干重 (g/m²)	茎干重 (g/m²)	千粒重 (g)	每株粒数
2014	CK	535.19±26.27 c	181.13±7.59 c	253.65±10.34 c	2.59±0.08a	3578.31±229.98 b
	E[CO_2]	632.71±33.48 ab	214.92±20.77 bc	320.73±19.03 b	2.75±0.12 a	4456.53±268.66 a
2015	CK	608.15±39.52 bc	256.64±4.42 ab	485.97±56.95 a	2.65±0.05 a	3559.94±363.98 b
	E[CO_2]	731.03±37.45 a	265.68±11.49 a	401.61±50.13 ab	2.71±0.10 a	3831.52±178.50 b
P 值	年份	0.22	0.00	0.00	0.85	0.11
	[CO_2]	0.00	0.04	0.83	0.13	0.01
	年份×[CO_2]	0.75	0.20	0.01	0.51	0.13

第四部分 大气 CO_2 浓度升高对北方主要 C_4 作物的影响

表 10-3 大气 CO_2 浓度升高对谷子各器官生长性状的影响

年份	处理	株高(cm)	穗长(cm)	茎粗(cm)	分蘖数量(个/株)	叶片数量(个/株)
2014	CK	109.72±2.63 a	17.67±0.56 a	0.57±0.02 a	1.42±0.05 b	12.16±0.22 a
	E[CO_2]	102.29±2.15 a	16.84±0.61 a	0.66±0.04 a	1.80±0.23 a	12.75±0.31 a
2015	CK	92.93±0.61 b	15.39±0.09 b	0.49±0.01 b	1.32±0.13 b	10.88±0.27 b
	E[CO_2]	90.95±4.68 b	16.93±0.48 a	0.55±0.03 a	1.51±0.11 ab	9.46±0.29 c
P 值	年份	0.00	0.01	0.00	0.87	0.11
	[CO_2]	0.13	0.35	0.00	0.01	0.00
	年份×[CO_2]	0.20	0.00	0.42	0.71	0.09

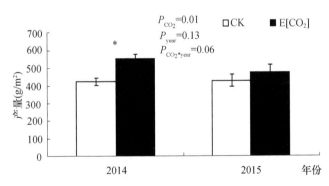

图 10-1 大气 CO_2 浓度升高对谷子产量的影响

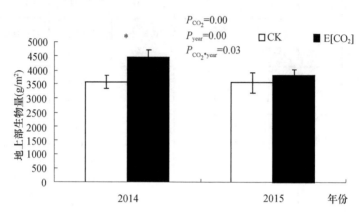

图 10-2 大气 CO_2 浓度升高对谷子地上部生物量积累的影响

10.3.2 P_n 与气体交换参数

大气 CO_2 浓度升高增加了所有谷子生长关键时期的 P_n ($P \leqslant 0.05$)。在大气 CO_2 浓度升高环境下,P_n 在 2014 年拔节、开花和灌浆期分别提高了 21%、73% 和 123%,在 2015 年开花期和灌浆期分别提高了 7% 和 19%。大气 CO_2 浓度升高条件下,G_s 在 2014 年(除抽穗期外)显著提高,但在 2015 年显著降低。T_r 的变化趋势与 G_s 一致。因而,大气 CO_2 浓度升高下,WUE_i 在 2014 年抽穗期、开花期和灌浆期分别显著提高了 77%、8% 和 41%,在 2015 年开花期和灌浆期分别显著提高了 45% 和 56%。对光响应曲线初始斜率的分析表明,谷子叶片在 CO_2 升高条件下没有出现光适应,P_n 在大气 CO_2 浓度为 600 μmol/mol 时没有达到饱和状态(表 10-4,图 10-3)。

第四部分 大气 CO_2 浓度升高对北方主要 C_4 作物的影响

表 10-4 大气 CO_2 浓度升高对谷子叶片气体交换特性的影响

年份	生育时期	处理	净光合速率 (P_n) [μmol/(m²·s)]	气孔导度 (G_s) [mol(H_2O)/(m²·s)]	蒸腾速率 (T_r) [m mol(H_2O)/(m²·s)]	水分利用效率 (WUE_i) [μmol(CO_2)/m mol(H_2O)]
2014	抽穗期	CK	20.62±0.94 b	0.17±0.01 a	2.74±0.14 a	132.40±4.49 c
		E[CO_2]	24.91±1.36 a	0.13±0.01 b	2.32±0.13 ab	209.65±7.21 a
	开花期	CK	11.62±0.58 c	0.10±0.00 c	2.10±0.09 b	116.75±2.13 cd
		E[CO_2]	20.20±0.71 b	0.16±0.01 a	2.77±0.12 a	125.89±3.38 c
	灌浆期	CK	10.44±0.61 c	0.07±0.00 d	1.46±0.08 c	155.59±2.19 b
		E[CO_2]	23.26±0.35 a	0.11±0.00 b	2.19±0.07 b	219.77±5.17 a
	P 值	生育时期	0.00	0.00	0.00	0.00
		[CO_2]	0.00	0.00	0.00	0.00
		交互	0.00	0.00	0.00	0.00

续表

年份	生育时期	处理	净光合率(P_n) [μmol/($m^2 \cdot s$)]	气孔导度(G_s) [mol(H_2O)/($m^2 \cdot s$)]	蒸腾速率(T_r) [mmol(H_2O)/($m^2 \cdot s$)]	水份利用效率(WUE$_i$) [μmol(CO_2)/m mol(H_2O)]
2015	开花期	CK	16.55±1.24 ab	0.12±0.01 a	3.90±0.37 a	138.17±7.00 b
		E[CO_2]	17.76±0.82 ab	0.09±0.01 b	3.77±0.27 a	200.19±12.36 a
	灌浆期	CK	15.86±1.01 b	0.12±0.01 a	2.37±0.21 b	136.88±10.74 b
		E[CO_2]	18.91±1.07 a	0.09±0.01 b	1.79±0.17 c	213.95±16.21 a
	P 值	生育时期	0.82	0.89	0.00	0.61
		[CO_2]	0.05	0.01	0.19	0.00
		交互	0.39	0.87	0.41	0.54

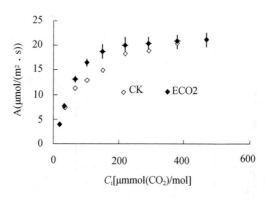

图 10-3 大气 CO_2 浓度升高对谷子叶片 A-C_i 特征曲线的影响

第四部分 大气 CO_2 浓度升高对北方主要 C_4 作物的影响

10.3.3 叶绿素荧光特性

大气 CO_2 浓度升高下,F_v/F_m 与 NPQ 在 2014 年和 2015 年均没有显著变化,Φ_{PSII}、qP 与 ETR 在 2014 年显著升高,但在 2015 年显著降低(表 10-5)。

表 10-5 大气 CO_2 浓度升高对谷子叶片叶绿素荧光特性的影响

(F_v/F_m－PSⅡ最大量子转化效率;F_v'/F_m'－PSⅡ实际量子转化效率;Φ_{PSII}－PSⅡ实际量子产量;qP－光化学淬灭系数;NPQ－非光化学淬灭系数)

年份	生育时期	处理	F_v/F_m	Φ_{PSII}	ETR	qP	NPQ
2014	抽穗期	CK	0.78±0.00 a	0.31±0.02 b	181.81±11.19 b	0.68±0.03 b	1.96±0.13 a
		E[CO_2]	0.78±0.20 a	0.37±0.02 a	223.39±9.73 a	0.79±0.02 a	1.87±0.18 a
	开花期	CK	0.75±0.00 c	0.21±0.02 d	122.92±9.31 d	0.49±0.04 cd	1.52±0.26 ab
		E[CO_2]	0.75±0.01 c	0.27±0.01 c	158.74±3.10 c	0.59±0.08 bc	1.64±0.20 ab
	灌浆期	CK	0.76±0.00 b	0.19±0.04 d	112.22±13.92 d	0.43±0.05 cd	1.37±0.08 bc
		E[CO_2]	0.75±0.00 c	0.27±0.01 c	160.93±8.32 c	0.60±0.03 bc	1.47±0.14 ab
	P 值	生育时期	0.00	0.00	0.00	0.00	0.03
		[CO_2]	0.81	0.00	0.00	0.02	0.74
		交互	0.11	0.46	0.46	0.60	0.56

续表

年份	生育时期	处理	F_v/F_m	Φ_{PSII}	ETR	qP	NPQ
2015	开花期	CK	0.75±0.01 ab	0.23±0.02 a	137.71±9.64 a	0.64±0.03 a	2.27±0.14 a
		E[CO_2]	0.74±0.00 bc	0.12±0.01 c	76.71±8.83 b	0.41±0.04 c	2.41±0.14 a
	灌浆期	CK	0.72±0.01 cd	0.23±0.01 a	135.09±8.47 a	0.56±0.03 b	1.65±0.11 b
		E[CO_2]	0.76±0.01 a	0.19±0.01 b	114.98±11.54 a	0.45±0.04 bc	1.63±0.09 b
	P值	生育时期	0.28	0.08	0.08	0.56	0.00
		[CO_2]	0.17	0.00	0.00	0.00	0.62
		交互	0.03	0.05	0.05	0.07	0.47

注：表中 F_v/F_m 为 PSII 最大光化学量子产量；Φ_{PSII} 为 PSII 实际光化学效率；ETR 为电子传递速率；qP 为光化学淬灭系数；NPQ 为非光化学淬灭系数。

10.3.4 基因表达

基于高通量测序分析结果显示，19 个基因表达上调，47 个基因表达下调。表 10-6 列出了不同表达模式基因的通路富集分析结果。大气 CO_2 浓度升高可显著影响植物激素信号转导、苯丙烷类生物合成、半乳糖代谢、鞘糖脂合成、角质、栓质与蜡质合成等合成和代谢通路的基因表达(Table 7)。高通量 RNA 测序结果可由反转录聚合酶链反应(RT-PCR)的结果证实(图 10-4)，表明高通量 RNA 测序结果的可靠性。

表 10-6 转录组基因表达的通路丰度分析

#	通路	P值	表达差异的基因
1	Plant hormone signal transduction	0.01	Si036015m.g, Si013648m.g, Si030198m.g, Si010073m.g, Si029090m.g, Si035574m.g

第四部分 大气 CO_2 浓度升高对北方主要 C_4 作物的影响

续表

#	通路	P 值	表达差异的基因
2	Phenylpropanoid biosynthesis	0.01	Si014344m.g,Si036462m.g,Si029879m.g,Si001092m.g
3	Galactose metabolism	0.01	Si029000m.g,Si035904m.g
4	Glycosphingolipid biosynthesis - globo series	0.03	Si035904m.g
5	Cutin, suberine and wax biosynthesis	0.03	Si006073m.g,Si019489m.g
6	Linoleic acid metabolism	0.06	Si013194m.g
7	Stilbenoid, diarylheptanoid and gingerol biosynthesis	0.07	Si014344m.g,Si029879m.g
8	Biosynthesis of secondary metabolites	0.07	Si006073m.g,Si037596m.g,Si014344m.g,Si019489m.g,Si036462m.g,Si029879m.g,Si001092m.g,Si022485m.g
9	Flavonoid biosynthesis	0.08	Si014344m.g,Si029879m.g
10	Phenylalanine metabolism	0.09	Si014344m.g,Si036462m.g
11	Sphingolipid metabolism	0.10	Si035904m.g
12	Biosynthesis of unsaturated fatty acids	0.11	Si035778m.g
13	Protein processing in endoplasmic reticulum	0.17	Si002731m.g,Si003151m.g
14	Glycerolipid metabolism	0.17	Si035904m.g
15	Inositol phosphate metabolism	0.17	Si000717m.g
16	Starch and sucrose metabolism	0.18	Si006474m.g,Si001092m.g
17	Cysteine and methionine metabolism	0.20	Si037596m.g
18	α-Linolenic acid metabolism	0.21	Si013194m.g

续表

#	通路	P 值	表达差异的基因
19	Phosphatidylinositol signaling system	0.21	Si000717m.g
20	Cyanoamino acid metabolism	0.23	Si001092m.g
21	Carotenoid biosynthesis	0.25	Si022485m.g
22	Circadian rhythm - plant	0.28	Si017194m.g
23	Metabolic pathways	0.32542	Si037596m.g, Si021485m.g, Si014344m.g, Si006474m.g, Si000717m.g, Si013194m.g, Si036462m.g, Si001092m.g, Si022485m.g
24	Pyrimidine metabolism	0.622732	Si021485m.g
25	Plant-pathogen interaction	0.920756	Si010446m.g
26	Carbon fixation in photosynthetic organisms		Si000774m.g
27	Glycolysis / Gluconeogenesis	0.00704	Si034948m.g

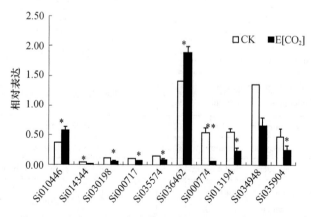

图 10-4　大气 CO_2 浓度升高对谷子叶片基因表达的影响

在木质素生物合成通路中，编码过氧化物酶的基因表达上调，但咖啡酰辅酶 A-O-甲基转移酶（CCoAOMT, caffeoyl-CoA O-methyl-transferase）基因表达下调。咖啡酰辅酶 A-O-甲基转移酶可催化芥子辅酶 A（sinapoyl-coA）的合成，过氧化物酶可催化松柏醇（coniferyl alcohol）形成愈创木基木质素（guaiacyl lignin）。在高 CO_2 浓度条件下，谷子中的参与植物激素信号转导引发气孔关闭的蛋白磷酸酶基因的表达下调，而富含亮氨酸的细胞壁强化蛋白的基因表达上调。在肌醇磷酸盐（inositol phosphate）代谢中，编码Ⅰ型肌醇多磷酸-5'-磷酸酶（IP5P1，Type Ⅰ inositol polyphosphate-5'-phosphatase）的基因在大气 CO_2 升高下表达下调。大气 CO_2 升高对糖基水解酶（glycosyl hydrolases）家族基因表达有副作用，而该家族可参与植物信号转导并导致茎尖萌发。大气 CO_2 浓度升高降低了脂氧合酶（lipoxygenase）的基因表达，该酶可催化 13-氧-10E-十二碳烯酸［13-OxODE，13-Oxo-10（E）-dodecenoic acid］的合成；而 α-半乳糖苷酶/α-N-乙酰半乳糖胺酶的基因表达显著降低。大气 CO_2 浓度升高显著抑制了 NADP-ME 的基因表达，而参与 3-磷酸甘油醛和 2-磷酸甘油醛转化的磷酸甘油酸变位酶（phosphoglycerate mutase）的基因表达显著下调。

10.4 讨论

本研究表明，谷子的籽粒产量与生物量在大气 CO_2 浓度升高条件下显著升高，是至今提升幅度最高的 C_4 植物，这与前人对 C_4 植物的研究结果不同。前人研究表明，C_4 植物对大气 CO_2 浓度升高环境的响应小于 C_3 植物（Manderscheid et al., 2014；Wand et al., 1999；Kadam et al., 2014）。大气 CO_2 浓度升高对干旱条件下 C_4 植物的胁迫影响有缓解作用（Leakey et al., 2006；Markelz et al., 2011；Kadam et al., 2014；Kooi et al., 2016）。可能是谷子具有较强抗旱性的原因，本研究中 2 年的谷子地上部生物量积累在大气 CO_2 浓度升高环境下显著增加，且与单位土地面积穗重和叶干重有关。

大气 CO_2 浓度升高下,分蘖数与穗粒数的增加可能是籽粒产量增加的主要贡献因素。大气 CO_2 浓度升高显著提高了谷子茎直径,降低了株高,这与前人对玉米(Xie et al.,2015)、高粱(Souza et al.,2010)形态适应的研究报道类似。但是谷子的生育期并未受到大气 CO_2 浓度升高的影响。Springer 和 Ward(2007)发现不同植物物种或种内单位的开花期对大气 CO_2 浓度升高的响应包括推迟、加速或没有变化,表明还需要更多的研究来探讨 CO_2 浓度升高对植物生育期的影响。作为 C_4 黍型作物,谷子具有高效的光合能力和较高的 WUE(Muthamilarasan et al.,2014)。这一 C_4 植物的进化策略可以提高叶片的 WUE 和 WUE_i,进而促进植株在大气 CO_2 浓度升高条件下的生长发育(Wang et al.,2015)。本研究表明,谷子在正常 CO_2 浓度与升高条件下均具有较高的水分利用效率(表10-4)。前人研究表明,大气 CO_2 浓度升高提升了不同水分条件下 C_4 植物的光合能力(Zong et al.,2014;Leakey et al.,2004)。另一方面,也有研究表明,大气 CO_2 浓度升高对离体或活体的光合能力、光合相关酶活性、生物量以及产量均没有显著影响(Leakey et al.,2006)。本研究中两年的数据表明,大气 CO_2 浓度升高下在叶片生长的各阶段,谷子的光合能力均显著提高(表10-4),尤其是 2014 年开花期与灌浆期显著提高了 26% 和 61%(刘紫娟 et al.,2017)。谷子的 A-C_i 曲线具有较高的 A-C_i 初始斜率,与其他 C_4 植物在大气 CO_2 浓度升高下接近光饱和的表现不同(图10-3)。大气 CO_2 浓度升高下,叶片发育过程中维持高光合能力很可能在生物量大幅增加的过程中起到关键作用。C_4 植物光合特性对水分胁迫的响应与 C_3 植物一样,在不同物种间存在较大差异。前人研究报道了 C_4 植物在水分胁迫条件下的光合抑制主要归因于气孔关闭,但也有人认为非气孔因素是主要原因(Ghannoum et al.,1998)。本研究中,在 2014 年谷子叶片的 G_s 与 T_r 在大气 CO_2 浓度升高条件下显著增加,相比于类似生境中的大多数 C_4 植物,这会有助于谷子的叶片温度降低,进而降低了热相关的危害。与前人研究一致,较低的 G_s 与较高的光合能力使谷子在 2015 年大气 CO_2 浓度升高环境下维持了比 C_3 植物更高的水分利用效率。

第四部分 大气 CO_2 浓度升高对北方主要 C_4 作物的影响

2014 年与 2015 年(表 10-5),谷子叶片的 ETR、qP 与 Φ_{PSII} 在大气 CO_2 浓度升高环境下均显著升高。前人对玉米的研究也表明,大气 CO_2 浓度升高促进了 ETR 和 Φ_{PSII} 的增加(Wang et al.,2015),这表明 ETR 和 Φ_{PSII} 的增加与光合能力增强关系密切。较大的 PSII 活性有助于产生足够的 NADPH 和 ATP 用于光合碳固定。但是,大气 CO_2 浓度升高抑制了 2015 年谷子的 ETR、qP 与 Φ_{PSII}。两年之间叶绿素荧光参数结果差异的原因仍需要被进一步研究。

转录组分析是揭示 C_4 植物逆境响应机制的有力工具。本研究中,我们运用转录组重测序技术鉴定了谷子中 66 种不同表达模式的基因。谷子适应大气 CO_2 浓度升高环境的相关基因表达数量不大,这可能与谷子对高 CO_2 环境的适应性较强有关。之前的研究表明谷子对贫瘠与干旱土壤的适应性较高,因此,被认为是气候变化时代的理想作物(Goron et al.,2015)。这一观点得到了生理与转录组结果的支持。大气 CO_2 浓度升高条件下,谷子的碳代谢改变更显著,可能是试图在细胞器水平调节器官间的碳分配。尤其明显的是木质素生物合成通路的辛烷酰辅酶 A 受抑制,而松柏醇酶过表达。前人研究也有类似结论,Körner 等(Körner et al.,2005)发现木质素含量在大气 CO_2 浓度升高条件下显著降低,表明碳向难溶性物质的分配比例降低,易溶性物质的比例可能会升高。

研究的结果进一步表明,大气 CO_2 浓度升高抑制了可增强茎分生的糖基水解酶家族基因的表达。同时,大气 CO_2 浓度升高上调了富含亮氨酸重复序列蛋白的基因表达,这有助于加强细胞壁。前人研究也表明了一些参与细胞壁组成的基因在 CO_2 浓度升高环境中的表达被上调(Li et al.,2008b;Ainsworth et al.,2006;Souza et al.,2010;Tian et al.,2010)。这些结果支持了本研究中形态适应、株高降低以及茎直径增大等结果很可能与氮代谢变化有关。谷子株高降低与茎机械强度增加会使茎在灌浆期的抗倒伏性能更强,进而增加了生物量积累与产量,这个结论与 Tian 等(2010)的结果一致。

气孔开闭在植物对大气 CO_2 浓度升高响应中的作用已被研究

人员广泛认可,然而,CO_2升高诱导气孔关闭的内在调节机制仍不明确。本研究中,大气 CO_2 升高下谷子叶片的气孔导度在 2014 年开花期与灌浆期均无显著变化(表 10-4)。编码可诱导气孔关闭的蛋白磷酸酶的基因表达下调也验证了这个结果。大气 CO_2 浓度升高对 α-半乳糖苷酶和 α-N-乙酰半乳糖胺酶的基因表达有负面影响,进而对半乳糖代谢、鞘糖脂生物合成、鞘脂代谢和甘油酯代谢产生抑制。在重力、光和盐胁迫下,植株体内肌醇磷酸代谢的信号通路常常增加(Morse et al.,1987;DeWald,2001)。在大气 CO_2 浓度升高条件下,与肌醇磷酸生物合成酶相关的转录物增强(Ainsworth et al.,2006)。但是,本研究表明,大气 CO_2 浓度增加抑制了谷子植株中的部分肌醇磷酸代谢。Khodakovskaya 等(2010)的研究表明,肌醇-(1,4,5)-三磷酸(InsP3)的基础水平降低,通过肌醇磷酸酶碳流增加可导致胞质 Pi 浓度增加,进而使光合固碳量增加。我们推测谷子在 CO_2 浓度增加下的响应机制与之类似。$NADP^+$ 依赖的苹果酸酶(NADP-ME)在 NADP-ME 亚型 C_4 植物中定位于维管束鞘细胞的叶绿体,发挥脱羧酶作用并释放 CO_2 供给光合作用过程(Wei et al.,2004)。本研究表明,大气 CO_2 升高抑制了谷子 NADP-ME 的基因表达,进而影响到其固碳过程。在 C_4 植物二羧酸循环中,$NADP^+$ 依赖的苹果酸酶催化苹果酸转化为丙酮酸。然而,本研究中谷子光合能力并没有降低,推测其中仍有其他方式为卡尔文循环提供碳源,这一现象的原因我们仍在进一步研究。

CO_2 浓度升高抑制了磷酸甘油酸变位酶的表达,表明糖酵解/糖异生受到抑制,导致 3-磷酸甘油醛降低,进而加强谷子的碳固定和叶片呼吸。糖酵解的抑制在小麦中也有报道(Buchner et al.,2015)。另一方面,大气 CO_2 浓度升高增加了大豆中糖酵解酶基因的转录丰度,3-磷酸甘油醛通过糖酵解通路转换次生代谢,特别是木质素、脂肪酸合成(Ainsworth et al.,2006)。我们发现谷子与其他 C_4 作物对 CO_2 浓度升高的响应特征具有很高的类似性,以上关于谷子的适应机制对于许多 C_4 植物适应气候变化是很重要的。

一些研究认为,通过碳代谢通路释放嘧啶氮在氮素再活化中具有重要作用(Buchner et al.,2015;Zrenner et al.,2009)。我们的研究表明,参与嘧啶代谢的 Si021485 基因的表达降低,最终会影响到氮代谢。

Rubisco 是光合作用与呼吸作用过程中的关键酶。前人研究表明,Rubisco 初始活性在大气 CO_2 浓度升高条件下降低,这也可能是 A_N 升高的原因之一(Jeroni et al.,2013)。CO_2 浓度升高条件下谷子 Rubisco 降低的现象尚需要进一步研究。

10.5 结论

本研究表明,尽管作为 C_4 植物,谷子在大气 CO_2 浓度升高下仍然会通过增加光合能力而增加生物量和籽粒产量。我们鉴定了一些对大气 CO_2 浓度升高敏感变化的基因,这些基因在细胞壁强度、茎分化、气孔导度、植物激素信号传导、碳固定、糖酵解/糖异生过程中发挥了重要作用。这些基因的改变使谷子株高降低、茎增粗,并增强了 CO_2 固定。这些结果表明在未来 CO_2 浓度升高环境中 C_4 作物谷子与 C_3 作物一样具有积极的正面响应。

第11章 大气 CO_2 浓度升高下青狗尾草与谷子光合特性比较研究

11.1 引言

目前,国内外对主要经济作物与树木在高浓度 CO_2 下的光合响应机制已做出了相关报道(Singh et al.,2014,Zong et al.,2014a,徐胜等,2015),但对于谷子及其杂草光合适应的研究却相对较少。青狗尾草(*Setaria viridis*)是谷子(*Setaria italica*)的主要伴生杂草,与谷子亲缘关系较近,被认为是谷子的野生近缘种(李伟等,2007),研究二者在高 CO_2 浓度下的适应机制不仅有助于阐明未来气候变化下二者的生长适应差异,也有助于对未来二者的竞争关系进行初步探索。

郝兴宇(2010)等通过对 FACE(Free Air CO_2 Enrichment)条件下谷子的农艺性状进行研究,证明高浓度 CO_2 有助于扩大谷子叶面积,促进株高和茎粗的增加,并提高水分利用效率。杨松涛(1997)等人对高浓度 CO_2 条件下谷子与狗尾草的叶片结构进行比较,证明大气 CO_2 浓度升高可以使谷子气孔密度和维管束鞘细胞叶绿体数量增加,狗尾草则反之。所以,高浓度大气 CO_2 可能会通过提高谷子叶片的光能吸收和转换能力以及叶肉细胞 CO_2 浓度而增强其光合性能,同时也降低了狗尾草的光合能力,但还需要进一步验证。因此,目前亟待通过在长期大气 CO_2 浓度升高条件下进行青狗尾草与谷子光合适应性的比较研究。

本研究通过研究青狗尾草与谷子叶片叶绿素荧光参数和光合气体交换参数在高浓度 CO_2 条件下的变化特征,阐明青狗尾草与谷子PSⅡ反应中心和气体交换对气候变化的响应调节方式,有助于理解二者对未来气候变化的光合作用适应机制,以期为气候变化背景下谷子与青狗尾草适应性的研究奠定基础。

第四部分 大气 CO_2 浓度升高对北方主要 C_4 作物的影响

11.2 试验设计与方法

11.2.1 试验设计

本试验在山西农业大学人工模拟控制 CO_2 浓度的开顶式生长室(OTC)进行。生长室长 6 m,宽 4 m,高 3 m,顶开口呈八角形。OTC 生长室内地面中部安装带小管辐射状的 CO_2 发散器,地面埋有与控制室 CO_2 钢瓶相连的通 CO_2 气体的管道,各生长室安装 CO_2 浓度传感器及水分测定器,可随时监测室内 CO_2 浓度,高 CO_2 浓度室内 CO_2 气体由减压阀控制的压缩 CO_2 钢瓶供给。处理期间的 CO_2 浓度由钢瓶流出,进出温室中由通气小管散开,风扇运行,保证室内 CO_2 浓度均匀,设置高 CO_2 浓度处理的目标浓度,检测仪连续监测,当室内 CO_2 浓度低于目标浓度时,自动将 CO_2 注入室内,使其浓度保持在 (560 ± 20) μmol/mol,通气时间 8:00—18:00,10 min 记录一次数据,正常 CO_2 浓度室内只通空气,不通入 CO_2。

采取池栽试验(池长 55 cm,宽 38 cm,深 30 cm),于 6 月 21 日播种,播前浇透水,播后覆土 1.5 cm。出苗后及时浇水,防止干旱,浇水在傍晚进行,并适时松土除草。幼苗长出第 4 片叶后间苗,每盆留 10 株,并施入 15 g $NH_4H_2PO_4$。

11.2.2 供试材料

供试材料为山西省太谷晋农种苗繁育场提供的谷子品种原平小谷,在华北地区春夏播种均可,株高 90~100 cm,生育期 90 d 左右。青狗尾草为山西农业大学农学院狗尾草资源圃提供,资源编号为 SX23(任彦鑫,2014),取自山西省忻州市。

11.2.3 指标测定与方法

气体交换参数测定:分别于拔节期、开花期和抽穗期使用便携式光合作用测定仪(LI-6400,Li-Cor Inc,Lincoln NE,USA),选取完全展开的大小均匀、无病虫害的倒一叶进行气体交换参数测定。测定项目包括净光合速率（P_n）、蒸腾速率（T_r）和气孔导度（G_s），并计算水分利用效率（WUE），WUE＝P_n/T_r。正常浓度 CO_2 和高浓度 CO_2 处理下的参比室 CO_2 浓度分别设为 380 μmol/mol 和 560 μmol/mol。测定时使用内置红蓝光源，光量子通量密度（PPFD）为 1600 μmol/(m²·s)，叶室温度设定在 25 ℃。

于开花期，选择大小均匀、无病虫害的谷子和青狗尾草倒一叶，使用 Li-6400 便携式光合作用测定仪于晴天 06:00,08:00,10:00,12:00,14:00,16:00,18:00 测定 P_n、G_s 和 T_r 的日进程变化，并计算 WUE。

叶片叶绿素荧光特性的测定：利用便携式调制叶绿素荧光仪（PAM-2500，Heinz Walz GmbH，Nuremberg，Germany）于晴天 9:00—11:30 测定倒一叶的叶绿素荧光参数。在叶片自然生长角度不变的情况下测定稳态荧光（F_s），同时记录叶表光强和叶温，随后加 1 个强闪光 [5000 μmol/(m²·s)，脉冲时间 0.7 s]，测定光下最大荧光（F_m'）；将叶片遮光，关闭作用光 5s 后暗适应 3 s，再打开远红光 5 s，测定光下最小荧光（F_0'）。叶片暗适应 30 min 后测定初始荧光（F_0），随后加 1 个强闪光（5000 μmol/(m²·s)，脉冲时间 0.7 s），测定最大荧光（F_m）。光系统Ⅱ（PSⅡ）最大光能转换效率 $F_v/F_m=(F_m-F_0)/F_m$；PSⅡ实际量子效率 $\Phi_{PSⅡ}=(F_m'-F_s)/F_m'$；光化学猝灭系数 qP＝$(F_m'-F_s)/(F_m'-F_0')$；非光化学猝灭系数 qN＝$1-(F_m'-F_0')/(F_m-F_0)=1-F_v'/F_v$。

地上部生物量的测定：于收获期收取地上部植株，并在鼓风式干燥箱中用 105℃杀青 15 min，70℃烘至恒重，用电子分析天平称取其生物量。

11.2.4　统计分析

以 Excel 进行数据处理，以 sigmaplot 图表绘制，并以 SAS 统计软件进行显著性分析。

第四部分　大气 CO_2 浓度升高对北方主要 C_4 作物的影响

11.3　结果分析

11.3.1　CO_2 浓度升高下青狗尾草与谷子气体交换参数的响应差异

大气 CO_2 浓度升高有助于改变青狗尾草与谷子叶片的光合气体交换特征。在拔节期和抽穗期，高浓度 CO_2 使青狗尾草 P_n 和 G_s 均显著提高，其中 P_n 较谷子高出 8.91% 和 8.45%，G_s 较谷子高出 102.55% 和 61.98%。同时，谷子的 T_r 显著下降，而青狗尾草未发生明显变化，所以谷子的 WUE 显著高于青狗尾草；在开花期，高浓度 CO_2 使谷子的 P_n 和 G_s 均持续增加，而青狗尾草显著下降。同时，高浓度 CO_2 使谷子的 T_r 显著高于青狗尾草，WUE 则无显著差异（图 11-1）。

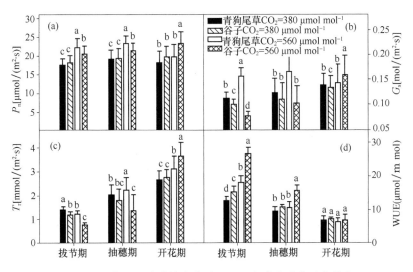

图 11-1　大气 CO_2 浓度升高条件下谷子与青狗尾草于拔节期、抽穗期和开花期的光合气体交换参数

注：图中相同小写字母表示在 0.05 水平不显著，下同。

11.3.2 CO_2浓度升高下青狗尾草与谷子气体交换参数日进程的响应差异

在抽穗期,大气CO_2浓度升高对谷子与青狗尾草光合日进程所产生的影响主要集中在10:00—16:00。在正常大气CO_2浓度下,青狗尾草旗叶的P_n在10:00—16:00显著低于谷子,大气CO_2浓度升高使青狗尾草的P_n得到显著提高,在10:00—16:00与谷子持平。在正常CO_2浓度下,谷子的G_s和T_r在10:00—16:00高于青狗尾草,CO_2浓度增加使谷子的G_s和T_r均显著降低,且与青狗尾草无显著差异。同时,CO_2浓度升高使谷子的WUE在8:00—12:00显著增强,青狗尾草则无明显变化(图11-2)。

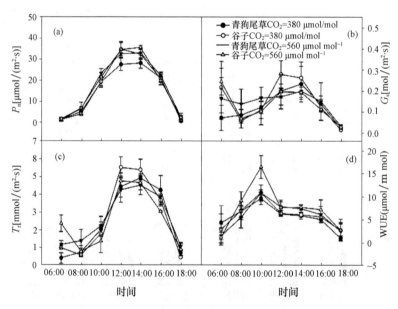

图11-2 大气CO_2浓度升高条件下抽穗期谷子与青狗尾草叶片光合气体交换参数日变化

11.3.3 CO_2浓度升高下青狗尾草与谷子叶绿素荧光参数的响应差异

在抽穗期,CO_2浓度升高使谷子的F_v/F_m和Φ_{PSII}均无显著提高,而使青狗尾草的Φ_{PSII}显著提高15.38%。在正常CO_2浓度下,青狗尾草的qP高于谷子,qN则无显著差异。大气CO_2浓度升高使青狗尾草的qP显著提高,提高幅度为6.1%,谷子则无明显变化。高浓度CO_2使青狗尾草的ETR提高15.79%,提升幅度高于谷子。所以,大气CO_2浓度升高条件下,青狗尾草的Φ_{PSII}、qP和ETR均较谷子得到显著提高(图11-3)。

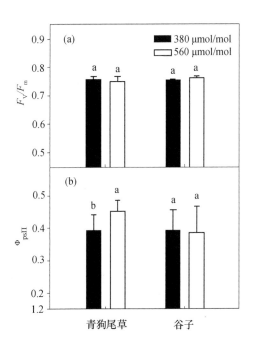

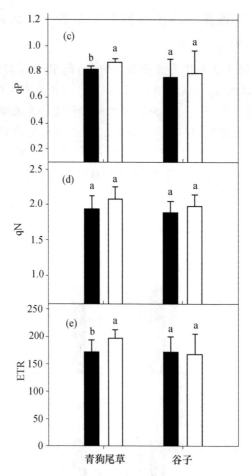

图 11-3 大气 CO_2 浓度升高条件下抽穗期谷子与青狗尾草叶绿素荧光特征

11.3.4 CO_2 浓度升高下青狗尾草与谷子产量构成因素的响应差异

大气 CO_2 浓度升高使谷子与青狗尾草的单穗重、穗粒重和地上部生物量均有显著提升，但对千粒重没有显著影响。其中，青狗尾草的单穗重、穗粒重和茎叶生物量在高浓度 CO_2 下分别提升了 22.37%、28.57% 和 14.75%，提升幅度均高于谷子(表 11-1)。

表 11-1　高浓度 CO_2 对谷子与狗尾草产量构成和生物量积累的影响(mean ± Std. Err)

	CO_2浓度 (μmol/mol)	单穗重 (g)	单穗粒重 (g)	千粒重 (g)	茎叶生物量 (g)
谷子	380	10.93±1.67 b	7.08±1.15 b	3.508±0.52 a	7.79±0.58 b
	560	13.07±2.01 a	8.30±1.39 a	3.762±0.59 a	8.38±0.71 a
增长幅度(%)		19.58	17.23	—	7.57
青狗尾草	380	0.22±0.01 b	0.14±0.04 b	0.52±0.03 a	2.44±0.40 b
	560	0.27±0.07 a	0.18±0.07 a	0.57±0.04 a	2.80±0.34 a
增长幅度(%)		22.73	28.57	—	14.75

注：同列不同小写字母表示差异显著($P<0.05$)，下同。

11.4　讨论

11.4.1　高浓度 CO_2 提高青狗尾草净光合速率

本研究中，在正常 CO_2 浓度下，青狗尾草的 P_n 在拔节期和抽穗期与谷子无差异，但高浓度 CO_2 使其显著提高。有研究表明，高浓度 CO_2 下狗尾草表皮细胞厚度显著增加，而谷子无明显变化(杨松涛等，1997)。本研究中高浓度 CO_2 有助于提高正午时刻青狗尾草的 P_n 值，所以青狗尾草叶片中较厚的表皮细胞可能在正午对叶绿体光合作用起到保护作用，有助于提高光合速率，缩小其与谷子干物质积累的差距。前人在关于大豆等多种豆科植物的研究中也发现，CO_2 浓度升高使其叶片栅栏组织层数增加，叶片加厚(Rogers et al.，1983，韩梅等，2006)。

同时，高浓度 CO_2 条件下，青狗尾草的 WUE 在拔节期和抽穗期显著低于谷子。这是因为高浓度 CO_2 显著升高了青狗尾草的 G_s,而对谷子的 G_s 影响不显著。有研究证明,高浓度 CO_2 下谷子叶片的气孔密度增加(杨松涛等,1997),也即气孔到达叶肉细胞的平均距离减小,此时由于受外界高 CO_2 浓度环境的影响,谷子叶片 C_i 已维持在较高水平,所以为了降低叶片蒸腾,必然要求表皮气孔具有较短的震荡过程与相对较低的 G_s,因而谷子具有较高的 WUE 而青狗尾草较低。

前人研究表明,大多数植株在高浓度 CO_2 下短期适应后光合能力上升而长期适应则显著下降,即光合下调现象,这种下调适应得到了生态学家的广泛关注(Lukac et al.,2010)。本研究中,青狗尾草 P_n 加强的维持期较短,在开花期显著降低,谷子的 P_n 则持续上升。这说明在生长后期,狗尾草出现较强的光合下调现象,有可能导致后期干物质积累减缓,而谷子的光合能力则仍维持在较高水平。

11.4.2　高浓度 CO_2 提升青狗尾草 PSⅡ 反应中心性能

本研究中在高 CO_2 浓度下,谷子与青狗尾草 PSⅡ 反应中心做出了不同响应。高浓度 CO_2 使青狗尾草 PSⅡ 反应中心的开放度提高,原初光能捕获率(Φ_{PSII})上升,天线色素光化学电子传递的份额(qP)上升,以及电子传递速率(ETR)上升,从而导致 PSⅡ 反应中心性能增强。而谷子叶片原初光能捕获率、PSⅡ 反应中心的光能转换、光化学粹灭与电子传递速率均未有明显变化,说明大气 CO_2 浓度升高下谷子 PSⅡ 反应中心性能未明显增强。张其德(1996)与 Zong 等(2014b)对大豆和玉米的研究也得出类似结论,高浓度 CO_2 下植株单位叶面积叶绿素含量增加,光能吸收和原初光能转化率增强,同时两个光系统之间激发能的分配格局发生改变。但也有研究得出不同的结论,卢从明等(1996)在水稻的研究中发现由于高浓度 CO_2 减少了单位叶面积叶绿素含量,从而限制了光吸收与 PSⅡ 反应中心性能的提高。不少研究表明,高浓度 CO_2 下植物光合作用的调节适应方向受氮限制现象影响较大(Reich et al.,2006a;Reich et al.,

2006b),因而这些不同的试验结果可能与供试土壤的营养有效性有关。

11.5 结论

本研究表明,大气 CO_2 浓度升高使青狗尾草与谷子的光合性能在拔节期和抽穗期均得到提升,其中青狗尾草的 P_n 值显著高于谷子。在抽穗期,高浓度 CO_2 下青狗尾草主要通过提高 PSⅡ反应中心原初光能捕获率、光化学粹灭系数、电子传递速率,并提高 12:00—14:00 的 P_n 值来增强光合性能,而谷子叶片 PSⅡ反应中心的光能转换、光化学粹灭和电子传递速率均未有明显变化;在开花期,高浓度 CO_2 下青狗尾草出现光合适应现象,谷子的光合能力则持续提高。在收获期,高 CO_2 浓度使谷子与青狗尾草茎叶生物量以及单穗重、单穗粒重均得到显著提升,且青狗尾草的提升幅度高于谷子。总之,CO_2 浓度升高不仅有助于提升青狗尾草的光合性能,而且有助于减少青狗尾草在干物质积累和产量构成方面与谷子的差距。

参考文献

韩梅,吉成均,左闻韵,等,2006. CO_2 浓度和温度升高对 11 种植物叶片解剖特征的影响[J]. 生态学报,26:326-33.

郝兴宇,李萍,林而达,等,2010. 大气 CO_2 浓度升高对谷子生长发育与光合生理的影响[J]. 核农学报,3:589-593.

李伟,智慧,王永芳,2007. 适合于狗尾草属遗传分析的 ISSR 标记筛选及反应体系优化研究[J]. 华北农学报,22:45-54.

刘紫娟,李萍,宗毓铮,等,2017. 大气 CO_2 浓度升高对谷子生长发育及玉米螟发生的影响[J]. 中国生态农业学报,1:55-60.

卢从明,张其德,唐崇钦,等,1996. CO_2 加富对杂交稻及其亲本原初光能转化效率和激发能分配的影响[J]. 生物物理学报,12:731-4.

任彦鑫,2014. 中国青狗尾草遗传多样性研究[D]. 太谷:山西农业大学.

山仑,邓西平,2000. 黄土高原半干旱地区的农业发展与高效用水[J]. 中国农业科技导报,2(4):34-38.

徐胜,陈玮,何兴元,等,2015. 高浓度 CO_2 对树木生理生态影响研究进展 [J]. 生态学报,35:2452-60.

杨松涛,胡玉熹,1997. CO_2 浓度倍增对 10 种禾本科植物叶片形态结构的影响 [J]. 植物学报:英文版,859-66.

张其德,卢从明,刘丽娜,1996. 二氧化碳加富对大豆叶片光系统 Ⅱ 功能的影响 [J]. 植物生态学报,20:517-23.

Ainsworth E A, Alistair R, Leakey A D B, et al, 2007. Does elevated atmospheric [CO_2] alter diurnal C uptake and the balance of C and N metabolites in growing and fully expanded soybean leaves? [J] Journal of Experimental Botany. (3):579-591.

Ainsworth E A, Long S P, 2005. What have we learned from 15 years of free-air CO_2 enrichment (FACE)? A meta-analytic review of the responses of photosynthesis, canopy properties and plant production to rising CO_2 [J]. New Phytologist, 165 (2):351-371.

Ainsworth E A, Rogers A, 2010. The response of photosynthesis and stomatal conductance to rising [CO_2]:mechanisms and environmental interactions [J]. Plant Cell and Environment, 30 (3):258-270.

Ainsworth E A, Rogers A, Vodkin L O, et al, 2006. The effects of elevated CO_2 concentration on soybean gene expression. An analysis of growing and mature leaves [J]. Plant Physiology, 142 (1):135-147.

Albert K R, Mikkelsen T N, Michelsen A, et al, 2011. Interactive effects of drought, elevated CO_2 and warming on photosynthetic capacity and photosystem performance in temperate heath plants [J]. Journal of Plant Physiology. 168 (13):1550-1561.

Allen L H, Kakani V G, Vu J C V, et al, 2011. Elevated CO_2 increases water use efficiency by sustaining photosynthesis of water-limited maize and sorghum [J]. Journal of Plant Physiology, 168 (16):1909-1918.

Aranjuelo I, Cabrera-Bosquet L, Morcuende R, et al, 2011. Does ear C sink strength contribute to overcoming photosynthetic acclimation of wheat plants exposed to elevated CO_2? [J]. Journal of Experimental Botany, 62 (11):3957-3969.

Audic S, Claverie J M, 1997. The significance of digital gene expression profiles [J]. Genome Research. 7:986-995.

Bennetzen J L, Schmutz J, Wang H, et al, 2012. Reference genome sequence of the model plant setaria [J]. Nature Biotechnology, 30: 555-561.

Bernacchi C J, Leakey A D B, Heady L E, et al, 2006. Hourly and seasonal variation in photosynthesis and stomatal conductance of soybean grown at future CO_2 and ozone concentrations for 3 years under fully open-air field conditions [J]. Plant, Cell and Environment, 29 (11): 2077-2090.

Buchner P, Tausz M, Ford R, et al, 2015. Expression patterns of C- and N-metabolism related genes in wheat are changed during senescence under elevated CO_2 in dry-land agriculture [J]. Plant Science, 236: 239-249.

Cabrera-Bosquet L, Albrizio R, Araus J L, et al, 2009. Photosynthetic capacity of field-grown durum wheat under different N availabilities: A comparative study from leaf to canopy [J]. Environmental and Experimental Botany, 67 (1): 145-152.

Chaves M M, Flexas J, Pinheiro C, 2008. Photosynthesis under drought and salt stress: regulation mechanisms from whole plant to cell [J]. Annals of Botany, 103 (4): 551-560.

Chun J A, Wang Q, Timlin D, et al, 2011. Effect of elevated carbon dioxide andwater stress on gas exchange and water use efficiency in corn [J]. Agricultural and Forest Meteorology, 151 (3): 378-384.

Cousins A B, Adam N R, Wall G W, et al, 2001. Rising CO_2-Future Ecosystems Ⅱ Reduced Photorespiration and Increased Energy-Use Efficiency in Young CO_2-Enriched Sorghum Leaves [J]. New Phytologist, 150 (2): 275-284.

Crous K Y, Walters M B, Ellsworth D S, 2008. Elevated CO_2 concentration affects leaf photosynthesis-nitrogen relationships in Pinus taeda over nine years in FACE [J]. Tree Physiology, 28 (4): 607-614.

Cseke L J, Tsai C J, Rogers A, , 2009. Transcriptomic comparison in the leaves of two aspen genotypes having similar carbon assimilation rates but different partitioning patterns under elevated [CO_2] [J]. New Phytologist, 182 (4): 891-911.

Dai H P, Zhang P P, Lu C, et al, 2011. Leaf senescence and reactive oxygen species metabolism of broomcorn millet (*Panicum miliaceum* L.) under drought condition [J]. Australian Journal of Crop Science, 5 (12): 1655-1660.

DeWald D B, 2001. Rapid accumulation of phosphatidylinositol 4,5-bisphosphate and inositol 1,4,5-trisphosphate correlates with calcium mobilization in salt-stressed arabidopsis [J]. Plant Physiology,126 (2):759-769.

Diao X,Schnable J,Bennetzen J L,et al,2014. Initiation of Setaria as a model plant [J]. Frontiers of Agricultural Science and Engineering,1(1):16-20.

Dosio G A A,Tardieu F,Turc O,2011. Floret initiation,tissue expansion and carbon availability at the meristem of the sunflower capitulum as affected by water or light deficits [J]. New Phytologist,189 (1):94-105.

Dyckmans J,Flessa H,2002. Influence of tree internal nitrogen reserves on the response of beech (*Fagus sylvatica*) trees to elevated atmospheric carbon dioxide concentration [J]. Tree Physiology,22 (1):41-49.

Ellsworth D S, 1999. CO_2 enrichment in a maturing pine forest: are CO_2 exchange and water status in the canopy affected? [J]. Plant, Cell and Environment,22:461-472.

Ellsworth D S, Reich P B, Naumburg E S, et al, 2004. Photosynthesis, carboxylation and leaf nitrogen responses of 16 species to elevated pCO_2 across four free-air CO_2 enrichment experiments in forest,grassland and desert [J]. Global Change Biology,10 (12):2121-2138.

Ephrath J E, Timlin D J, Reddy V, et al, 2011. Irrigation and elevated carbon dioxide effects onwhole canopy photosynthesis and water use efficiency in cotton (*Gossypium hirsutum* L.) [J]. Plant Biosystems,145 (1):202-215.

Farooq M, Wahid A, Kobayashi N, et al, 2009. Plant drought stress: effects, mechanisms and management [J]. Agronomy for Sustainable Development,29 (1):185-212.

Farquhar G D,Sharkey T D,1982. Stomatal conductance and photosynthesis [J]. Annual Review of Plant Physiology,33:317-345.

Franzaring J, Weller S, Schmid I, et al, 2011. Growth, senescence and water use efficiency of spring oilseed rape (*Brassica napus* L. cv. Mozart) grown in a factorial combination of nitrogen supply and elevated CO_2 [J]. Environmental and Experimental Botany,72 (2):284-296.

Fukayama H,Sugino M,Fukuda T,et al,2011. Gene expression profiling of rice grown in free air CO_2 enrichment (FACE) and elevated soil temperature [J]. Field Crop Research. 121 (1):195-199.

Furbank R T, Chitty J A, vonCaemmerer S, et al, 1996. Antisense RNA inhibition of *RbcS* gene expression reduces rubisco level and photosynthesis in the C_4 plant *Flaveria bidentis* [J]. Plant Physiology, 111 (3): 725-734.

Galle A, Feller U, 2007. Changes of photosynthetic traits in beech saplings (*Fagus sylvatica*) under severe drought stress and during recovery [J]. Physiologia plantarum, 131 (3): 412-421.

Galle A, Florez-Sarasa I, Tomas M, et al, 2009. The role of mesophyll conductance during water stress and recovery in tobacco (*Nicotiana sylvestris*): acclimation or limitation? [J]. Journal of Experimental Botany, 60 (8): 2379-2390.

Galloway J N, Townsend A R, Erisman J W, et al, 2008. Transformation of the nitrogen cycle: Recent trends, questions, and potential solutions [J]. Science, 320 (5878): 889-892.

Gao J, Han X, Seneweera S, et al, 2015. Leaf photosynthesis and yield components of mung bean under fully open-air elevated [CO_2] [J]. Journal of Integrative Agriculture, 20 (5): 977-983.

Gesch R W, Kang I H, Gallo-Meagher M, et al, 2003. Rubisco expression in rice leaves is related to genotypic variation of photosynthesis under elevated growth CO_2 and temperature [J]. Plant, Cell and Environment. 26 (12): 1941-1950.

Ghannoum O, 2009. C_4 photosynthesis and water stress. Annals of Botany. 103 (4): 635-644.

Ghannoum O, Conroy J P, 1998a. Nitrogen deficiency precludes a growth response to CO_2 enrichment in C_3 and C_4 Panicum grasses [J]. Australian Journal of Plant Physiology. 25 (5): 627-636.

Ghannoum O, Siebke K, Von C S, et al, 1998b. The photosynthesis of young Panicum C_4 leaves is not C_3-like [J]. Plant, Cell and Environment, 21: 1123-1131.

Ghannoum O, Caemmerer S V, Ziska L H, et al, 2000a. The growth response of C_4 plants to rising atmospheric CO_2 partial pressure: a reassessment [J]. Plant, Cell and Environment, 23 (9): 931-942.

Ghannoum O, Von Caemmerer S, Ziska L H, et al, 2000b. The growth response of C_4 plants to rising atmospheric CO_2 partial pressure: a reassessment [J]. Plant, Cell and Environment, 23 (9): 931-942.

Ghannoum O, Evans J R, Chow W S, et al, 2005. Faster rubisco is the key to superior nitrogen-use efficiency in NADP-malic enzyme relative to NAD-malic enzyme C_4 grasses [J]. Plant Physiology,137 (2):638-650.

Ghannoum O, Evans J R, Caemmerer S V, 2011. Nitrogen and Water Use Efficiency of C_4 Plants. In:Raghavendra AS,Sage RF (eds) C_4 Photosynthesis and Related CO_2 Concentrating Mechanisms [M]. Springer, Dordrecht, 129-146.

Goron T L,Raizada M N,2015. Genetic diversity and genomic resources available for the small millet crops to accelerate a New Green Revolution [J]. Frontiers in Plant Science,1 (6):157.

Gupta P, Duplessis S, White H, et al, 2010. Gene expression patterns of trembling aspen trees following long-term exposure to interacting elevated CO_2 and tropospheric O_3[J]. New Phytologist,167 (1):129-142.

Hao X Y,Han X,Lam S K,et al,2012. Effects of fully open-air [CO_2] elevation on leaf ultrastructure,photosynthesis,and yield of two soybean cultivars [J]. Photosynthetica,50 (3):362-370.

Hao X Y,Li P,Feng Y X, 2013. Effects of fully open-air CO_2 elevation on leaf photosynthesis and ultrastructure of Isatis indigotica Fort. PLoS One:e74600.

Hu L X,Wang Z L,Huang B R,2010. Diffusion limitations and metabolic factors associated with inhibition and recovery of photosynthesis from drought stress in a C_3 perennial grass species [J]. Physiologia Plantarum,139 (1):93-106.

Hunt H V, Campana M G, Lawes M C, et al, 2011. Genetic diversity and phylogeography of broomcorn millet (*Panicum miliaceum* L.) across Eurasia [J]. Molecular Ecology,20 (22):4756-4771.

IPCC,2013. Summary for Policymakers. Climate Change: The Physical Science Basis. Contribution of Working Group I to the Fifth Assessment Report of the Intergovernmental Panel on Climate Change [J]. Cambridge University Press, Cambridge,United Kingdom and New York,NY,USA.

Jenny R,Watling M C,Press W,et al, 2000. Elevated CO_2 Induces Biochemical and Ultrastructural Changes in Leaves of the C_4 Cereal Sorghum [J]. Plant Physiology,123:1143-1152.

Jeroni,Galmés,Iker,et al,2013. Variation in Rubisco content and activity under variable climatic factors [J]. Photosynth Research,117:73-90.

Jia G,Huang X,Zhi H,et al,2013. A haplotype map of genomic variations and genome-wide association studies of agronomic traits in foxtail millet (Setaria italica) [J]. Nature Genetics, 45 (8):957.

Jin J, Armstrong R, Tang C, 2019. Impact of elevated CO_2 on grain nutrient concentration varies with crops and soils-A long-term FACE study [J]. Science of The Total Environment, 651:2641-2647.

Körner C, Asshoff R, Bignucolo O, et al, 2005. Carbon flux and growth in mature deciduous forest trees exposed to elevated CO_2 [J]. Science, 309 (5739):1360-1362.

Kadam N T, Xiao G, Jean M R, et al, 2014. Agronomic and physiological responses to high temperature drought: and elevated CO_2 in cereals [J]. Advances in Agronomy, 127: 111-156.

Khalil K, Dahal K P, Badawi M A, et al, 2013. Long-term growth under elevated CO_2 suppresses biotic stress genes in non-acclimated, but not cold-acclimated winter wheat [J]. Plant and Cell Physiology (11):1751-1768.

Khodakovskaya M, Sword C, Wu Q, et al, 2010. Increasing inositol (1, 4, 5)-trisphosphate metabolism affects drought tolerance, carbohydrate metabolism and phosphate-sensitive biomass increases in tomato [J]. Plant Biotechnology Journal, 8 (2):170-183.

Kim H Y, Lim S S, Kwak J H, et al, 2011. Dry matter and nitrogen accumulation and partitioning in rice (Oryza sativa L.) exposed to experimental warming with elevated CO_2 [J]. Plant and Soil, 342 (1-2):59-71.

Kimball B A, 2016. Crop responses to elevated CO_2 and interactions with H_2O, N, and temperature [J]. Current Opinion in Plant Biology, 31:36-43.

Kontunen-Soppela S, Riikonen J, Ruhanen H, et al, 2010. Differential gene expression in senescing leaves of two silver birch genotypes in response to elevated CO_2 and tropospheric ozone [J]. Plant, Cell and Environment, 33 (6):1016-1028.

Kooi C J, Reich M, Löw M, et al, 2016. Growth and yield stimulation under elevated CO_2 and drought: a meta-analysis on crops [J]. Environmental and Experimental Botany, 122:150-157.

Lata C, Gupta S, Prasad M, 2013. Foxtail millet: a model crop for genetic and genomic studies in bioenergy grasses [J]. Critical Reviews in Biotechnology, 33 (3):328-343.

Lattanzi F A, 2005. The sources of carbon and nitrogen supplying leaf growth:

assessment of the role of stores with compartmental models [J]. Plant Physiology,137 (1):383-395.

Leakey A D B, 2009. Rising atmospheric carbon dioxide concentration and the future of C₄ crops for food and fuel [J]. Proceedings of the Royal Society B: Biological Sciences,276 (1666).

Leakey A D, Bernacchi C J, Ort D R, et al, 2010. Long-term growth of soybean at elevated [CO₂] does not cause acclimation of stomatal conductance under fully open-air conditions [J]. Plant,Cell and Environment,29 (9):1794-1800.

Leakey A D B, 2009. Rising atmospheric carbon dioxide concentration and the future of C₄ crops for food and fuel [J]. Proceedings of the Royal Society B-Biological,276 (1666):2333-2343.

Leakey A D B,Bernacchi C J,Dohleman F G,et al,2004. Will photosynthesis of maize (*Zea mays*) in the US Corn Belt increase in future [CO₂] rich atmospheres? An analysis of diurnal courses of CO₂ uptake under free-air concentration enrichment (FACE) [J]. Global Change Biology, 10 (6): 951-962.

Leakey A D B, Uribelarrea M, Ainsworth E A, et al, 2006. Photosynthesis, productivity,and yield of maize are not affected by open-air elevation of CO₂ concentration in the absence of drought [J]. Plant Physiology, 140 (2): 779-790.

Leakey A D B, Xu F, Gillespie K M, et al, 2009. Genomic basis for stimulated respiration by plants growing under elevated carbon dioxide [J]. Proceedings of the National Academy of Sciences of the United States of America. 106 (9):3597-3502.

Li P A E,Andrew A,Leakey D B,et al,2008a. Arabidopsis transcript and metabolite profiles:ecotype-specific responses to open-air elevated [CO₂] [J]. Plant,Cell and Environment,31:1673-1687.

Li P, Ainsworth E A, Leakey A D B, et al, 2008b. Arabidopsis transcript and metabolite profiles:Ecotype-specific responses to open-air elevated [CO₂][J]. Plant,Cell and Environment,31 (11):1673-1687.

Li P,Sioson A,Mane S P,et al,2006. Response diversity of Arabidopsis thaliana ecotypes in elevated [CO₂] in the field [J]. Plant Molecular Biology,62 (4-5):593-609.

Liu Z, Sun N, Yang S, et al, 2013. Evolutionary transition from C_3 to C_4 photosynthesis and the route to C_4 rice [J]. Biologia,68 (4):577-586.

Long S P, Ainsworth E A, Rogers A, et al, 2004. Rising atmospheric carbon dioxide: Plants face the future [J]. Annual Review of Plant Biology, 55: 591-628.

Long S P, Bernacchi C J, 2003. Gas exchange measurements, what can they tell us about the underlying limitations to photosynthesis? Procedures and sources of error [J]. Journal of Experimental Botany,54 (392):2393-2401.

Lukac M, Calfapietra C, Lagomarsino A, et al, 2010. Global climate change and tree nutrition: effects of elevated CO_2 and temperature [J]. Tree Physiology 30,1209-1220.

Luo Y, Su B, Currie W S, et al, 2004. Progressive nitrogen limitation of ecosystem responses to rising atmospheric carbon dioxide [J]. Bioscience,54 (8):731-739.

Manderscheid R, Erbs M, Weigel HJ, 2014. Interactive effects of free-air CO_2 enrichment and drought stress on maize growth [J]. European Journal of Agronomy. 52:11-21.

Markelz R J, Strellner R S, Leakey A D, 2011. Impairment of C_4 photosynthesis by drought is exacerbated by limiting nitrogen and ameliorated by elevated [CO_2] in maize [J]. Journal of Experimental Botany,62:3235-3246.

Matthew M, C, Kimball B A, Brooks T J, et al, 2001. CO_2 enrichment increases water-use efficiency in sorghum [J]. New Phytologist,151 (2):407-412.

Moore B D, Shu-Hua C, Edwards G E, 1986. The influence of leaf development on the expression of C_4 metabolism in *Flaveria trinervia*, a C_4 dicot [J]. Plant and Cell Physiology,6:1159-1167.

Morse M J, Crain R C, Satter R L, 1987. Light-stimulated inositol phospholipid turnover in *Samanea saman* leaf pulvini [J]. Proceedings of the National Academy of Sciences of the United States of America,84 (20):7075-7078.

Mortazavi A, Williams B A, McCue K, et al, 2008. Mapping and quantifying mammalian transcriptomes by RNA-Seq [J]. Nature Methods. 5:621-628.

Mu Q Z, Zhao M S, Heinsch F A, et al, 2007. Evaluating water stress controls on primary production in biogeochemical and remote sensing based models [J]. Journal of Geophysical Research-Biogeosciences. 112 (G1).

Muthamilarasan M, Khandelwal R, Yadav C B, et al, 2014. Identification and molecular characterization of MYB transcription factor super-family in C_4 model plant foxtail millet (Setaria italica L.). PLoS One. 9:e109920.

Nure F, Md A K, Romel A, et al, 2010. Water stress effects on growth of dipterocarpus turbinatus seedlings[J]. Forest Science and Technology,6 (1): 18-23.

Pinheiro C,Passarinho J A,Ricardo C P,2004. Effect of drought and rewatering on the metabolism of Lupinus albus organs [J]. Journal of Plant Physiology. 161 (11):1203-1210.

Prieto J A,Louarn G,Perez PeÑA J,et al,2012. A leaf gas exchange model that accounts for intra-canopy variability by considering leaf nitrogen content and local acclimation to radiation in grapevine (Vitis vinifera L.) [J]. Plant,Cell & Environment,35 (7):1313-28.

Reich P B,Hungate B A,Luo Y,2006. Carbon-nitrogen interactions in terrestrial ecosystems in response to rising atmospheric carbon dioxide [J]. Annual Review of Ecology,Evolution,and Systematics,37 (1):611-636.

Reynolds J F, Kemp P R, Ogle K, et al, 2004. Modifying the'pulse-reserve' paradigm for deserts of North America:precipitation pulses,soil water,and plant responses [J]. Oecologia. 141 (2):194-210.

Robredo A,Pérez-López U,Miranda-Apodaca J,et al,2011. Elevated CO_2 reduces the drought effect on nitrogen metabolism in barley plants during drought and subsequent recovery [J]. Environmental and Experimental Botany. 71:399-408.

Robredo A,Perez-Lopez U,Lacuesta M,et al,2010. Influence of water stress on photosynthetic characteristics in barley plants under ambient and elevated CO_2 concentrations [J]. Biology Plantarum,54 (2):285-292.

Rogers H H,Thomas J F,Bingham G E,1983. Response of agronomic and forest species to elevated atmospheric carbon dioxide [J]. Science,220,428-429.

Salvucci M E,Crafts-Brandner S J,2004. Mechanism for deactivation of Rubisco under moderate heat stress [J]. Physiologia Plantarum,122 (4):513-519.

Saxe H,Ellsworth D S,Heath J,1998. Tree and forest functioning in an enriched CO_2 atmosphere [J]. New Phytologist,139:395-436.

Schneider M K, Luscher A, Richter M, et al, 2004. Ten years of free-air CO_2 enrichment altered the mobilization of N from soil in Lolium perenne

L. swards [J]. Global Change Biology,10 (8):1377-1388.

Sekhar K M,Kota V R,Reddy T P,et al,2020. Amelioration of plant responses to drought under elevated CO_2 by rejuvenating photosynthesis and nitrogen use efficiency:implications for future climate-resilient crops [J]. Photosynth Research.

Seneweera S, Makino A, Hirotsu N, et al, 2011. New insight into photosynthetic acclimation to elevated CO_2: The role of leaf nitrogen and ribulose-1, 5-bisphosphate carboxylase/oxygenase content in rice leaves [J]. Environmental and Experimental Botany. 71 (2):128-136.

Seneweera S, Makino A, Mae T, et al, 2005. Response of rice to p (CO_2) enrichment:the relationship between photosynthesis and nitrogen metabolism [J]. Journal of Crop Improvement. 38 (1):289-299.

Seneweera S P,Ghannoum O,Conroy J P,et al,2002. Changes in source sink relations during development influence photosynthetic acclimation of rice to free air CO_2 enrichment (FACE) [J]. Functional Plant Biology,29:945-953.

Shi K,Li X,Zhang H,et al,2015. Guard cell hydrogen peroxide and nitric oxide mediate elevated CO_2-induced stomatal movement in tomato [J]. New Phytologist,208:342-353.

Singh S,Reddy V,2014. Combined effects of phosphorus nutrition and elevated carbon dioxide concentration on chlorophyll fluorescence, photosynthesis, and nutrient efficiency of cotton [J]. Journal of Plant Nutrition & Soil Science,177:892-902.

Soba D,Shu T,Runion G B,et al,2020. Effects of elevated [CO_2] on photosynthesis and seed yield parameters in two soybean genotypes with contrasting water use efficiency [J]. Environmental and Experimental Botany,178:104154.

Solomon S, 2007. Climate change 2007: the physical science basis: contribution of Working Group I to the Fourth Assessment Report of the Intergovernmental Panel on Climate Change [R]. Cambridge,UK and NY,USA:Cambridge Univ Pr.

Souza A P D, Gaspar M, Silva E A D, et al, 2010. Elevated CO_2 increases photosynthesis, biomass and productivity, and modifies gene expression in sugarcane [J]. Plant,Cell and Environment,31 (8):1116-1127.

Springer C J,Ward J K,2007. Flowering time and elevated atmospheric CO_2 [J]. New Phytologist,176:243-255.

Steffen W, Sanderson A, Tyson P D, et al, 2003. Global change and the earth system [M]. In:Eaeja (ed) A Planet Under Pressure. Springer, Berlin.

Stitt M, 1991. Rising CO_2 levels and their potential significance for carbon flow in photosynthetic cells [J]. Plant, Cell & Environment, 14 (8):741-762.

Swemmer A M, Knapp A K, Snyman H A, 2007. Intra-seasonal precipitation patterns and above-ground productivity in three perennial grasslands [J]. Journal of Ecology, 95 (4):780-788.

Taiz L, Zeiger E, 2006. Plant Physiology [M]. Fourth edn. Sinauer Associates, Inc., Sunderland.

Takatani N, Ito T, Kiba T, et al, 2014. Effects of high CO_2 on growth and metabolism of arabidopsis seedlings during growth with a constantly limited supply of nitrogen [J]. Plant and Cell Physiology, 55 (2):281-292.

Tausz-Posch B, Dempsey R, Norton R, et al, 2013. The effect of elevated CO_2 on photochemistry and antioxidative defence capacity in wheat depends on environmental growing conditions-A FACE study [J]. Environmental and Experimental Botany, 88:81-92.

Taylor G, Street N R, Tricker P J, et al, 2005. The transcriptome of Populus in elevated CO_2 [J]. New Phytologist, 167 (1):143-154.

Thomson A M, Brown R A, Rosenberg N J, et al, 2005. Climate change impacts for the conterminous USA: an integrated assessment. Part 3. Dryland production of grain and forage crops [J]. Climate Change, 69:43-65.

Tian B, Wang J, Zhang L, et al, 2010. Assessment of resistance to lodging of landrace and improved cultivars in foxtail millet [J]. Euphytica, 172 (3):295-302.

Tobita H, Uemura A, Kitao M, et al, 2011. Effects of elevated atmospheric carbon dioxide, soil nutrients and water conditions on photosynthetic and growth responses of*Alnus hirsuta* [J]. Functional Plant Biology, 38 (9):702-710.

Underhill AP, 1997. Current issues in Chinese Neolithic archaeology [J]. Journal of World Prehistory, 11 (2):103-160.

Varone L, Ribas-Carbo M, Cardona C, et al, 2012. Stomatal and non-stomatal limitations to photosynthesis in seedlings and saplings of Mediterranean speciespre-conditioned and aged in nurseries: Different response to water stress [J]. Environmental and Experimental Botany, 75:235-247.

von Caemmerer S, Furbank R, 1999. Modeling C_4 photosynthesis [M]. In: Sage

RF, Monson RK (eds) C_4 plant biology. Academic Pr, San Diego, CA, USA, 173-211.

Vu JCV, Allen LH, 2009. Growth at elevated CO_2 delays the adverse effects of drought stress on leaf photosynthesis of the C_4 sugarcane [J]. Journal of Plant Physiology, 166 (2): 107-116.

Wall G W, Brooks T J, Adam N R, et al, 2001. Elevated atmospheric CO_2 improved Sorghum plant water status by ameliorating the adverse effects of drought [J]. New Phytologist, 152: 231-248.

Wand S J E, Midgley G F, Jones M H, et al, 1999. Responses of wild C_4 and C_3 grass (*Poaceae*) species to elevated atmospheric CO_2 concentration: a meta-analytic test of current theories and perceptions [J]. Global Change Biology. 5 (6): 723-741.

Wang M J, Xie B Z, Fu Y M, et al, 2015. Effects of different elevated CO_2 concentrations on chlorophyll contents, gas exchange, water use efficiency and PSII activity on C_3 and C_4 cereal crops in a closed artificial ecosystem [J]. Photosynth Research, 126: 351-362.

Wang X Z, Taub D R, 2010. Interactive effects of elevated carbon dioxide and environmental stresses on root mass fraction in plants: a meta-analytical synthesis using pairwise techniques [J]. Oecologia, 163 (1): 1-11.

Watanabe M, Watanabe Y, Kitaoka S, et al, 2011. Growth and photosynthetic traits of hybrid larch F1 (*Larix gmelinii* var. *japonicax L-kaempferi*) under elevated CO_2 concentration with low nutrient availability [J]. Tree Physiology, 31 (9): 965-975.

Watling J R, Press M C, 1997. How is the relationship between the C_4 cereal Sorghum bicolor and the C_3 root hemi-parasites Striga hermonthica and Striga asiatica affected by elevated CO_2? [J]. Plant, Cell and Environment, 20: 1292-1300.

Wei C, Jing S Z, Fang Z, et al, 2004. Photosynthetic features of transgenic rice expressing sorghum C_4 type NADP-ME [J]. Acta Botanica Sinica, 46 (7): 873-882.

Xie H C, Liu K Q, Sun D D, et al, 2015. A field experiment with elevated atmospheric CO_2-mediated changes to C_4 crop-herbivore interactions [J]. Scientific Reports, 5: 13923.

Xu Z, Zhou G, Shimizu H, 2009. Are plant growth and photosynthesis limited by pre-drought following rewatering in grass? [J]. Journal of Experimental Botany, 60 (13): 3737-3749.

Xu Z Z, Zhou G S, Wang Y H, 2007. Combined effects of elevated CO_2 and soil drought on carbon and nitrogen allocation of the desert shrub Caragana intermedia [J]. Plant and Soil, 301 (1-2): 87-97.

Yahdjian L, Sala O E, 2006. Vegetation structure constrains primary production response to water availability in the Patagonian steppe [J]. Ecology, 87 (4): 952-962.

Zanetti S, Hartwig U A, Luscher A, et al, 1996. Stimulation of symbiotic N_2 fixation in *Trifolium repens* L under elevated atmospheric pCO_2 in a grassland ecosystem [J]. Plant Physiology, 112 (2): 575-583.

Zhou Z C, Gan Z T, Shangguan Z P, et al, 2009. China's Grain for Green Program has reduced soil erosion in the upper reaches of the Yangtze River and the middle reaches of the Yellow River [J]. International Journal of Sustainable Development and World Ecology, 16 (4): 234-239.

Zhou Z C, Shangguan Z P, 2009. Effects of Elevated CO_2 Concentration on the Biomasses and Nitrogen Concentrations in the Organs of Sainfoin (*Onobrychis viciaefolia* Scop.) [J]. Agricultural Sciences in China, 8 (4): 424-430.

Ziska L H, Bunce J A, 1997. Influence of increasing carbon dioxide concentration on the photosynthetic and growth stimulation of selected C_4 crops and weeds [J]. Photosynthesis Research, 54: 199-208.

Zong Y, Wang W, Xue Q, et al, 2014. Interactive effects of elevated CO_2 and drought on photosynthetic capacity and PS II performance in maize [J]. Photosynthetica, 52: 63-70.

Zong Y Z, Shangguan Z P, 2016. Increased sink capacity enhances C and N assimilation under drought and elevated CO_2 conditions in maize [J]. Journal of Integrative Agriculture, 15 (012): 2775-2785.

Zrenner R, Riegler H, Marquard C R, et al, 2009. A functional analysis of the pyrimidine catabolic pathway in Arabidopsis [J]. New Phytologist (2): 117-132.